MW00713358

Janus
Life Science

GLOBE FEARON EDUCATIONAL PUBLISHER
A Division of Simon & Schuster
Upper Saddle River, New Jersey

CONTRIBUTORS

Mary K. Friedland
Winifred Ho Roderman
Susan Echaore-Yoon
Nancy Lobb
Gerald Booth

CONSULTANTS

William A. Jensen, Ph.D., University of California, Berkeley
Gilbert Yee, Fremont, California
Katharine Barrett, Lawrence Hall of Science, University of California, Berkeley
John Bartevian, Mt. Eden High School, Hayward, California
Myra Lappin, M.D., San Francisco, California
Susan B. Sherman, San Francisco State University, San Francisco, California

REVIEWER

Peggy A. Prazma, Piedmont High School, San Jose, California

ARTISTS

Ellen Beier, Gwen Sanders, Lisa Haderlie Baker, Margaret Sanfilippo, Nancy Kirk

PHOTO CREDITS

cover: FPG, 3: Grant Heilman, 49: : Dennis Demello, N.Y. Zoological Society, 77: Photo Researchers, 77: Grant Heilman, 77: Hal H. Harrison, Photo Researchers, 77: Grant Heilman, 95: Barbara Rios, Photo Researchers, 95: Kathy Sloane, Photo Researchers

ISBN 0-835-91388-0

Printed in the United States of America
5 6 7 8 9 10 99

GLOBE FEARON EDUCATIONAL PUBLISHER
A Division of Simon & Schuster
Upper Saddle River, New Jersey

Contents

Introduction

In this book, you will learn about some of the things that scientists study. You will learn some of the things that scientists have discovered. You will explore and discover facts the same way scientists do, by experimenting, observing, and recording. And you will learn scientific information about the world that you can use right away.

Living Things

Life science is the study of living things. You are a living thing. And you need other living things to stay alive.

What is a living thing? How are you like other living things? How are you different? What living things do you need in order to stay alive?

Life scientists study living things. They discover ways living things are the same and ways they are different. Some scientists study certain animals, such as monkeys or insects. Other scientists study another animal—humans—to find out how our bodies work or how we live.

We know a lot about plants, ourselves, and other animals because of what scientists have learned. We know how to raise animals and plants for food. We also know how our bodies use food to keep us healthy. And we know that we need plants and animals in order to stay alive.

In this book, you'll learn some of the facts that scientists have discovered about living things. As you study living things, you will use the same methods that scientists use: You will observe living things and record what you find out. And you will learn some of the ways living things help each other to stay alive.

GREEN PLANTS

What kind of living things are green plants? How are they like other living things? What do green plants need to stay alive? In this section, you'll learn many facts about these living things. And you'll learn how green plants play an important part in our lives.

Contents

Introduction

Picture this:

You're moving into your own place. It's your first apartment! You've got a bed. You've got a few chairs and a table. You've got a stereo and some other things.

You put all your things in the apartment. But even with all your things, the apartment looks a little empty. What else can you put in the apartment?

You don't have much money. But you want to get something that will make the place look great. What can you get?

You can get some green plants. You can buy them or you can start them yourself. But either way, you'll have to take care of those plants. What will you do to take care of them? How will you keep them green? How will you keep them growing? How will you keep them alive?

In this section, you'll learn about one type of living thing: green plants. You'll learn how to start your own plants. You'll learn what plants need to grow and stay healthy. And you'll learn some scientific facts about plants.

You'll know a lot about green plants. So, when you move into your own place, you can use plants to make it look great.

Unit 1

Getting Started

Think of a plant. Is that plant green?

There are many different kinds of plants in this world. But most of the plants we see are green, like the plant you probably thought of.

You'll learn about green plants in this book. How? By watching plants grow.

Many green plants can be started from seeds. So that's how you'll begin—by starting plants from seeds.

- What makes seeds grow?
- How do seeds become plants?

You'll find out in this unit.

Before You Start

You'll be using the science words below. Find out what they mean. Look them up in the Glossary that's at the back of this book. On a separate piece of paper, write what the words mean.

1. **seedling**
2. **soil**
3. **sprout**

Getting Seeds to Grow

You'll be starting plants from three kinds of seeds: radish seeds, bean seeds, and tomato seeds.

The radish seeds will come from a packet of seeds. The bean seeds will come from the grocery store. And the tomato seeds will come right out of a tomato.

You'll do three things to get those seeds to grow. What do you think those three things are?

These are the three things you'll do:

1. You'll plant the seeds in soil.
2. You'll give the seeds water.
3. You'll put the planted seeds in a warm place.

Look around the room. Find a warm place where you can put the planted seeds. Where is that place?

You can get your bean seeds and radish seeds to sprout quickly. Here's how: Soak them in water for a while *before* you plant them in soil.

Soak the seeds right now. You'll need two cups and some water.

- Fill each cup half full with water.
- Put eight lima beans in one cup.
- Put eight radish seeds in the other cup.
- Let the seeds soak overnight.

Experiment 1

How long do seeds take to sprout?

Some seeds take longer to sprout than others. For example, your radish, bean, and tomato seeds will sprout on different days. How many days will each take? Do this experiment and find out.

Materials (What you need)

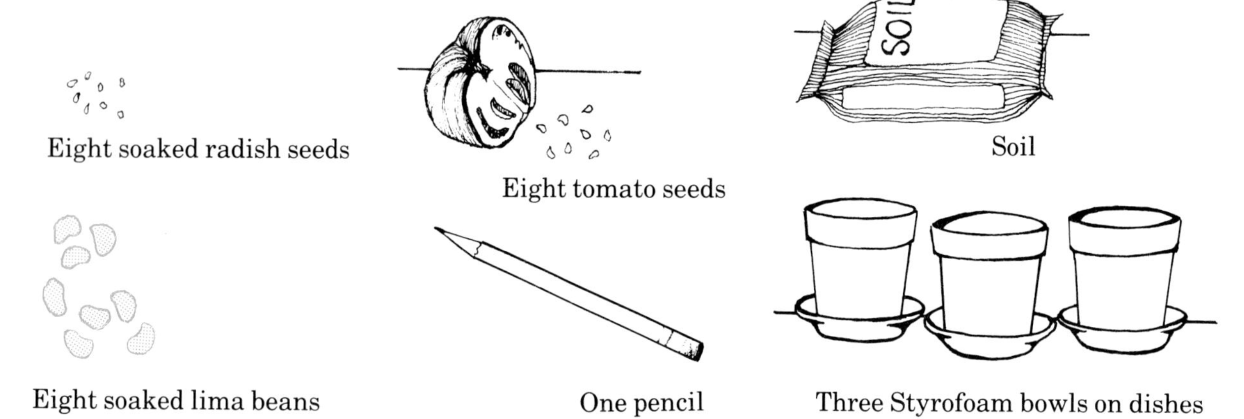

Procedure (What you do)

1. Write your name on each bowl. Then write today's date on each bowl. Next, write *radish* on the first bowl, *bean* on the second bowl, and *tomato* on the third bowl.

2. Turn the bowls upside down. With your pencil, poke three holes near the bottom of each bowl.

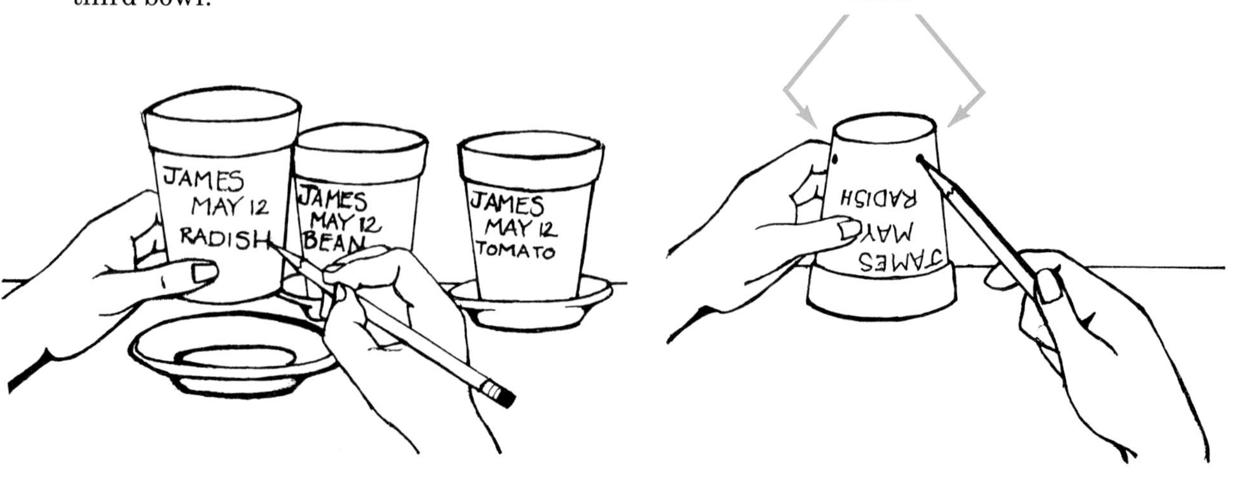

3. Fill each bowl almost to the top with soil. Put the seeds in the right bowls. Spread the seeds out. Cover them with a thin layer of soil.

4. Pour water on the top of the soil. Move the bowls to a warm place. Look at them every day. As you watch them, you will be able to answer the questions below.

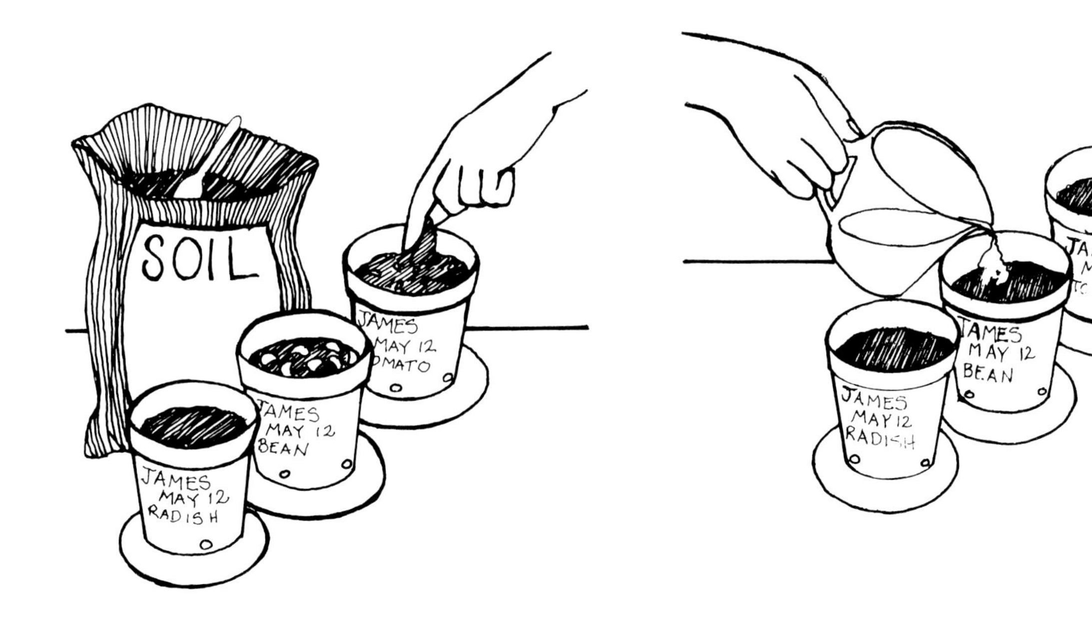

Observations (What you see)

1. When did the first radish seedling come up?
2. When did the first bean seedling come up?
3. When did the first tomato seedling come up?

Conclusions (What you learn)

1. How many days did the first radish seed take to sprout?
2. How many days did the first bean seed take to sprout?
3. How many days did the first tomato seed take to sprout?

Plant Watch

Watch your radish seeds grow into seedlings. See how they change.

Two days after you've planted the seeds, start watching them. Watch them every day for four days. Keep a record of what you see.

A Record of How Radish Seeds Grow

Two Days After Planting

Take a seed out of the soil. Look at it. What do you see growing out of the seed? On a separate piece of paper, draw what you see.

Three Days After Planting

Take another seed out of the soil. What's growing out of the seed? On a separate piece of paper, draw what you see.

Four Days After Planting

By now, you should have a seedling growing out of the soil. Carefully take it out of the soil.

What does the seedling look like? Describe it. On a separate piece of paper, draw what you see.

Five Days After Planting

Take another seedling out of the soil. Look at it.

Now look at the seedling you drew yesterday. How is today's seedling different? On a separate piece of paper, draw a picture of the seedling.

Three Main Parts

Look at the radish seedling you have removed from the soil. What are the parts that make it up?

One part of the seedling was in the soil. This part is the **root**. What do you think the root does for the plant?

A second part of the seedling spread out above the soil. This part is made up of **leaves**. What do you think leaves do?

A third part of the seedling supported the leaves and connected them to the roots. This part is the **stem**.

Most green plants have the same main parts you see in your radish seedling.

Find the drawing you made of the radish seedling. Write *root*, *stem*, and *leaves* next to those parts in the drawing. Your drawing should now look like the picture on this page.

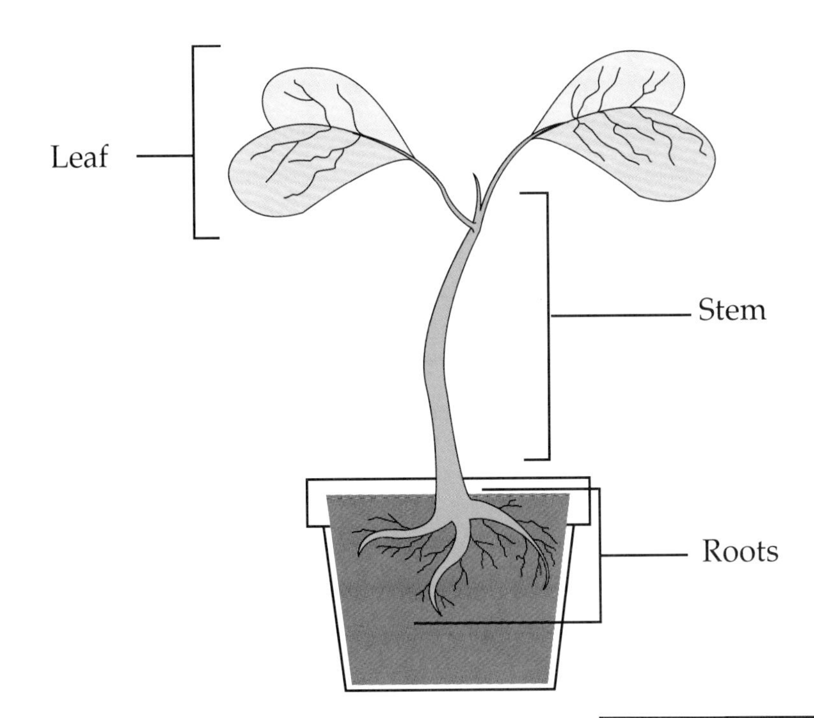

You will need other seedlings to do the experiments that come later. Plant at least 4 bean seeds and 4 tomato seeds now. Follow the directions for Experiment 1, but put each bean seed in a separate cup, put 2 of the tomato seeds in separate cups, and put 2 tomato seeds in one cup.

Review

Use the words you learned on page 6 to answer the questions.

1. a. What do you call the earth that plants grow in?
   b. What do you call a very young plant?
   c. What word means "start to grow"?

Look back at page 7 to answer these questions.

2. a. How did you get bean and radish seeds to sprout quickly?
   b. What three things did you do to get your seeds to grow?

Check These Out

1. Make a Science Notebook for this section. Use it to keep a record of what you learn about green plants. Put your list of glossary words and their meanings in it. Also keep your notes from experiments and the Plant Watch sections in it. You can keep anything else you learn about green plants in your notebook too.
2. Find out how to **transplant** seedlings. Then transplant the seedlings that you have grown. Put each seedling in its own cup.
3. Take your radish seedlings home. Plant them in a large pot or in the ground. In three weeks, some radishes should be ready to eat.
4. Make a seed collection. Get seeds from fruits, vegetables, nuts, or outdoor plants. Tape the seeds on a large piece of heavy paper. Below each seed, write the name of the plant that the seed came from.
5. As you work through this section, you may want to find out more about plants. You can do this by looking in an encyclopedia or by getting books from a library. You can also talk to an expert, such as a gardener, florist, or botanist.

   Here are some things you may want to find out:
   - What's inside a seed? What does *germinate* mean?
   - Seeds often end up far away from the plants they came from. Seeds travel in many different ways. What are those ways?

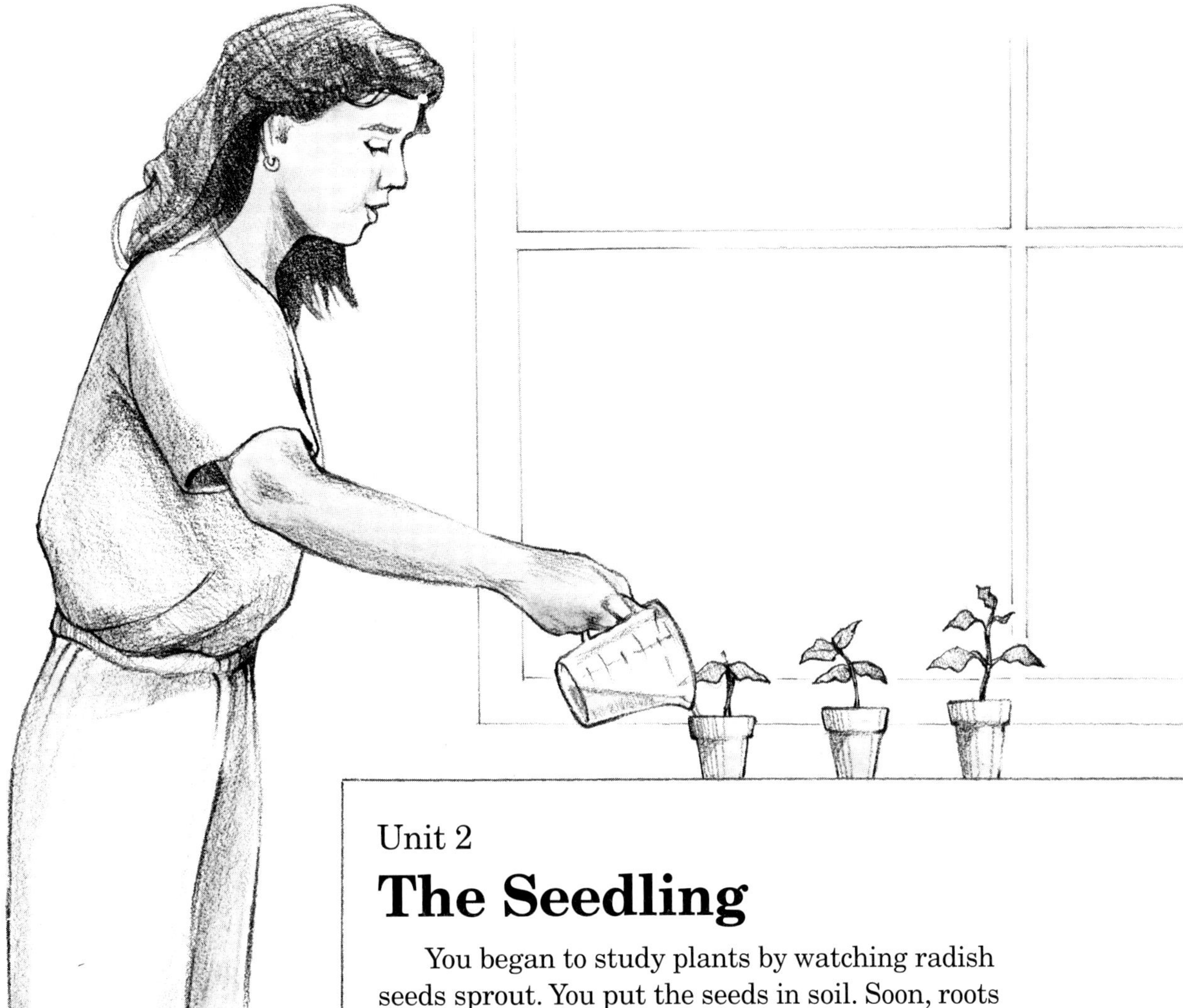

Unit 2

The Seedling

You began to study plants by watching radish seeds sprout. You put the seeds in soil. Soon, roots pushed out from the seeds. The roots grew quickly. Then a stem and leaves appeared on top of the soil. Your seeds had grown into seedlings.

- What do the parts of a seedling look like close-up?
- What things do plants need so they can grow?

You'll find the answers in this unit.

Before You Start

You'll be using the science words below. Find out what they mean. Look them up in the Glossary. On a separate piece of paper, write what the words mean.

1. **dissolve**
2. **fertilizer**
3. **nutrient**

Close-Up of a Seedling

Scientists keep a record of what they study. When they study plants, they look at the plants very carefully and write down what they see. When they look at the same plants later, they can tell if there have been any changes by looking at their records.

Choose a tomato seedling that you have grown. It should have at least two leaves. Carefully take the seedling out of the soil. Then brush the loose soil off the root.

Look for the three main parts. Do you see a root, a stem, and some leaves?

Describe the seedling. How many leaves does it have? What color are they? What do the leaves feel like? Are they smooth, rough, or fuzzy?

What color is the stem? Is it straight or curved?

What does the root look like? What color is it? Is it wet or dry? Does it have just one part or does it have many parts?

Measure the length of each part with a ruler.

On a separate sheet of paper, draw the seedling. Show the colors of the parts. Show the length of each part. Show how the surface of each part looks close-up.

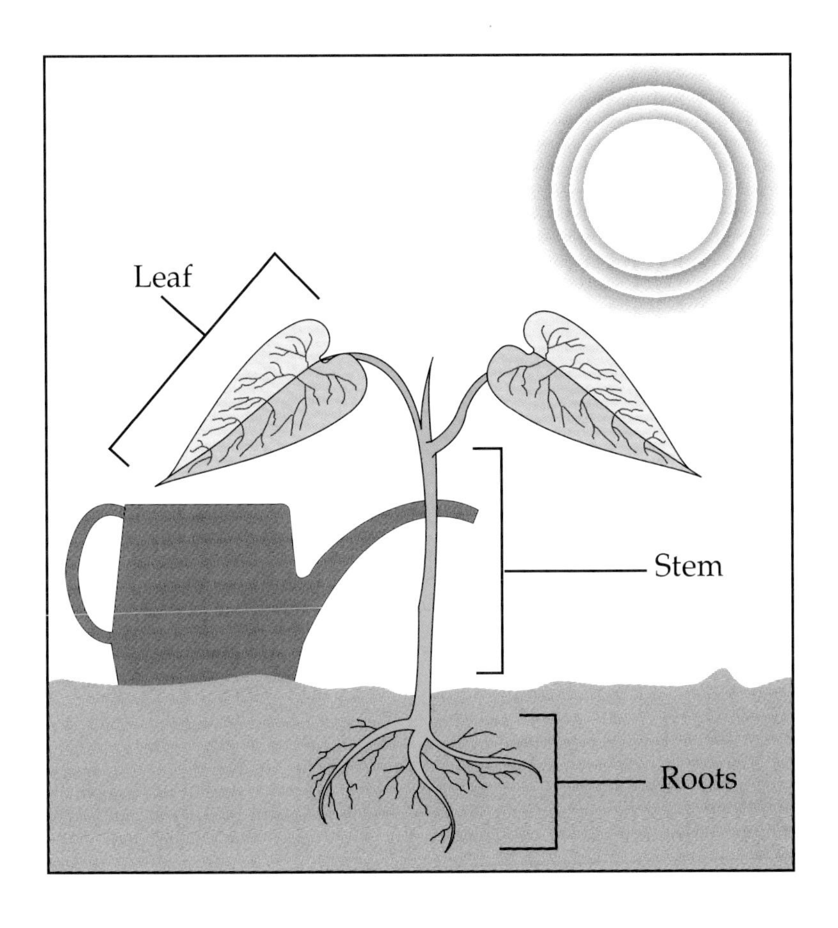

Plants Make Their Own Food

How are plants different from animals? A green plant is different from an animal in the way it gets food. Animals take in food from outside their bodies. They eat plants and other animals. Plants, however, make food inside their bodies. They don't need to eat.

But even though plants don't eat, they still need things from their surroundings so they can live and grow.

What things besides food do you need to live? Does a plant need any of the things you need? Do you think a plant needs water? Air? Is there anything a plant needs that you don't need?

Think about the main parts of a plant. What do you think a plant gets through its leaves? What does it get through its roots?

Plants Need Water

When you planted your seeds, you watered them. Your seeds grew into seedlings, and you're still watering them.

What happens to the water you give your seedlings? Get a bean seedling and find out.

Cut off the stem close to the soil. Look at the stump—the part that's left in the soil. What is coming out of the stump?

Hold the cut-off seedling straight up and down. Squeeze the bottom of the stem. What comes out?

Crush a leaf between your fingers. What comes out?

Something wet came out of the stump, the stem, and the leaves. That something is water. When you watered the soil, that water went into the seedling.

A plant, like your seedling, is made up mostly of water. Water helps the plant keep its shape. You can see this for yourself.

Carefully take a tomato seedling out of the soil. Lay it on a table for a day.

The next day, pick up the seedling. Look at it. What does the seedling look like?

A B

Plants Need Nutrients

Imagine this: You're growing a healthy green plant. You give it enough water. But after a few months, its leaves turn yellow and some fall off the plant. What's wrong?

The plant might need **nutrients**. You can give the plant nutrients by adding **fertilizer** to the soil. With the nutrients in the fertilizer, the plant can become healthy again.

So what are the nutrients in plant fertilizer? The nutrients are simple chemicals. Plants need them to help build their bodies and to help run their food-making machinery. These nutrients are not the same as food because plants do not get energy from them.

There are three main plant nutrients: nitrogen, phosphorus, and potassium. These chemicals are part of most soil. Water in the soil **dissolves** them and carries them into the roots of plants.

But plants can use up the nutrients in their soil. That's when they need the added nutrients in fertilizer.

The pictures on this page show two house plants. They sprouted from seeds at the same time. They are in the same kind of soil. They grew in the same place. The plant in picture **A** got fertilizer. The plant in picture **B** did not get fertilizer.

How are the plants different? What do you think is the cause of this difference?

Plants Need Light

People say plants "like" the sun. You'll see what they mean if you look at a plant that's by a window. The plant stem tilts toward the sun. Its leaves face the sun, as if to soak up the sunlight.

If you watch that plant for a day, you'll see that the stem tip moves. In the morning, the tip points east, toward the sun. At noon, when the sun is straight above, the tip points straight up. During the afternoon, the tip follows the sun to the west.

The plant is seeking light. It needs light to make food.

If you watch the plant for several days, you'll see that it is growing. The stem gets longer, and new leaves come out. You'll also see that the plant is growing in a certain direction.

Stems of green plants grow toward one thing. What is that thing? Do this experiment and find out.

Experiment 2

What do plant stems grow toward?

Materials

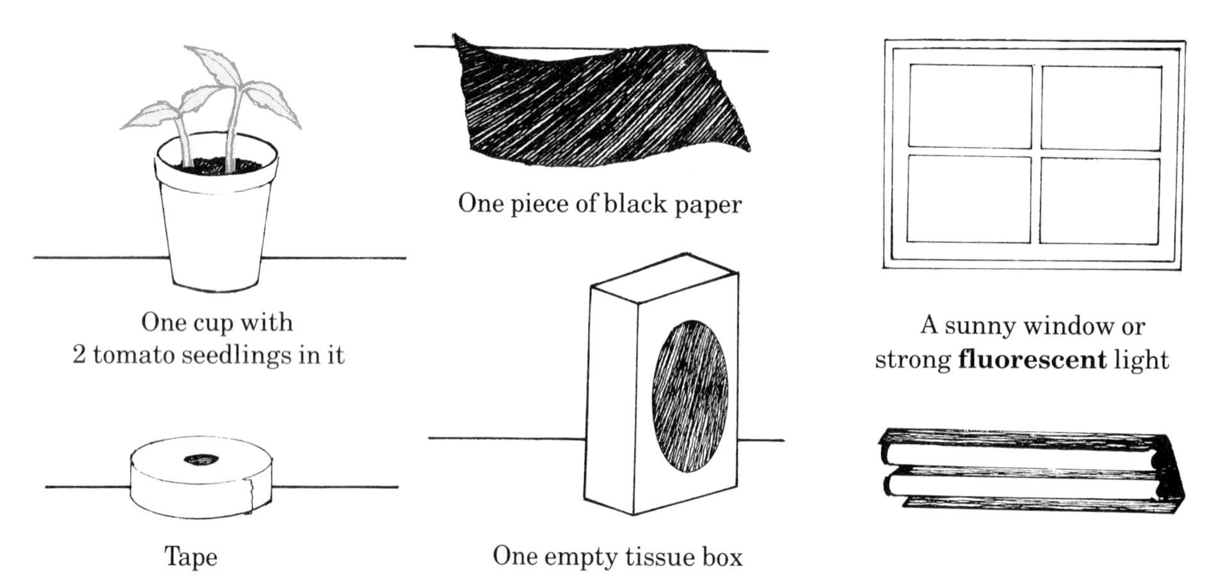

One cup with 2 tomato seedlings in it

One piece of black paper

A sunny window or strong **fluorescent** light

Tape

One empty tissue box

Procedure

1. Stand the tissue box on its end.

2. Put the seedlings inside the box.

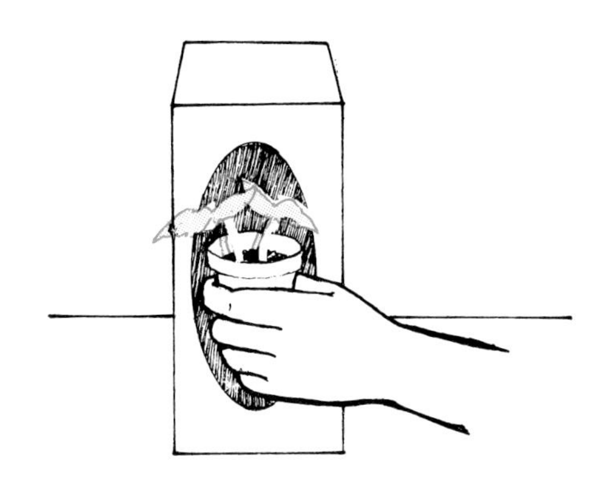

3. Tape black paper over most of the opening on the box. Leave two inches at the bottom uncovered.

4. Put the box near a window or fluorescent light. Wait three days.

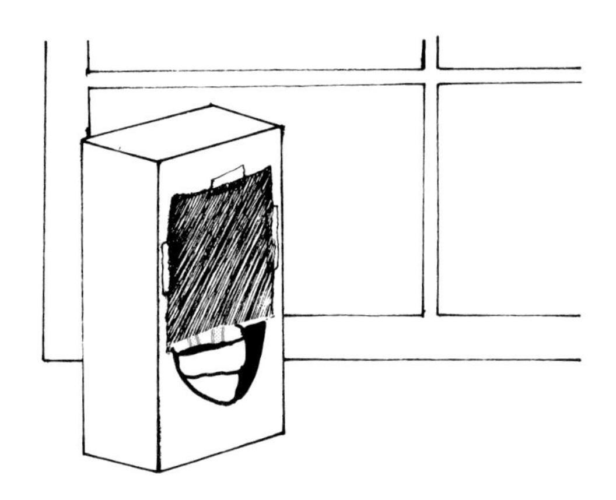

Observations

After three days, take the black paper off the tissue box. Look at the seedlings.

Which way are the stems pointed? Straight up? Toward the opening of the box? Toward the back of the box?

Conclusions

What do plant stems grow toward?

Plant Watch

In Experiment 2 on pages 18–19, you learned that plant stems grow toward light. As the light moves, the plant stem bends in the direction of the light.

Keep a record of what you learned. Get the seedlings you used in Experiment 2. On a separate piece of paper, draw the cup and the seedlings.

Show how the seedlings grew toward the light. In your picture, show which way the light is coming from. Show which way the stems bend. Show which way the leaves are facing.

Review

Show what you learned in this unit. Match the words on the left with the clues on the right.

1. water	a. This part of the plant takes in water.
2. root	b. This is sometimes given to plants that look sick.
3. leaves	c. This helps a plant keep its shape.
4. nutrients	d. Minerals in the soil do this in water.
5. dissolve	e. Plants need these to stay healthy.
6. light	f. These parts of a plant face the sun.
7. fertilizer	g. Plant stems grow toward this.
8. stem	h. The tip of this part of the plant moves.

Check These Out

1. Find out about fertilizers. Visit a plant store or plant nursery. Ask these questions: What's in fertilizers? Why are there different kinds of fertilizers?
2. What is compost? Find out. Ask someone who grows plants. Or look in books on how to grow plants.
3. Choose a plant to learn about. It can be any kind of plant. Find out these things:
   - How often should you water it?
   - How much light does it need?
   - When should it be fertilized?

   Tell the class what you learned.
4. Here are more things you may want to find out:
   - What is plant classification? How do scientists classify plants?
   - Who was Gregor Mendel and what did he discover about plants? What are genes? What are chromosomes?
   - What is a Venus's flytrap? What other plants are like it?

Unit 3

Roots

You've learned that a plant has three main parts. It has a root, a stem, and leaves. All three parts work together to make food.

You'll find out what each part does. Let's begin by looking at roots.

- What do roots do?
- How do roots work?

You'll learn the answers in this unit.

Before You Start

You'll be using the science words below. Find out what they mean. Look them up in the Glossary. On a separate piece of paper, write what the words mean.

1. **absorb**
2. **stored**

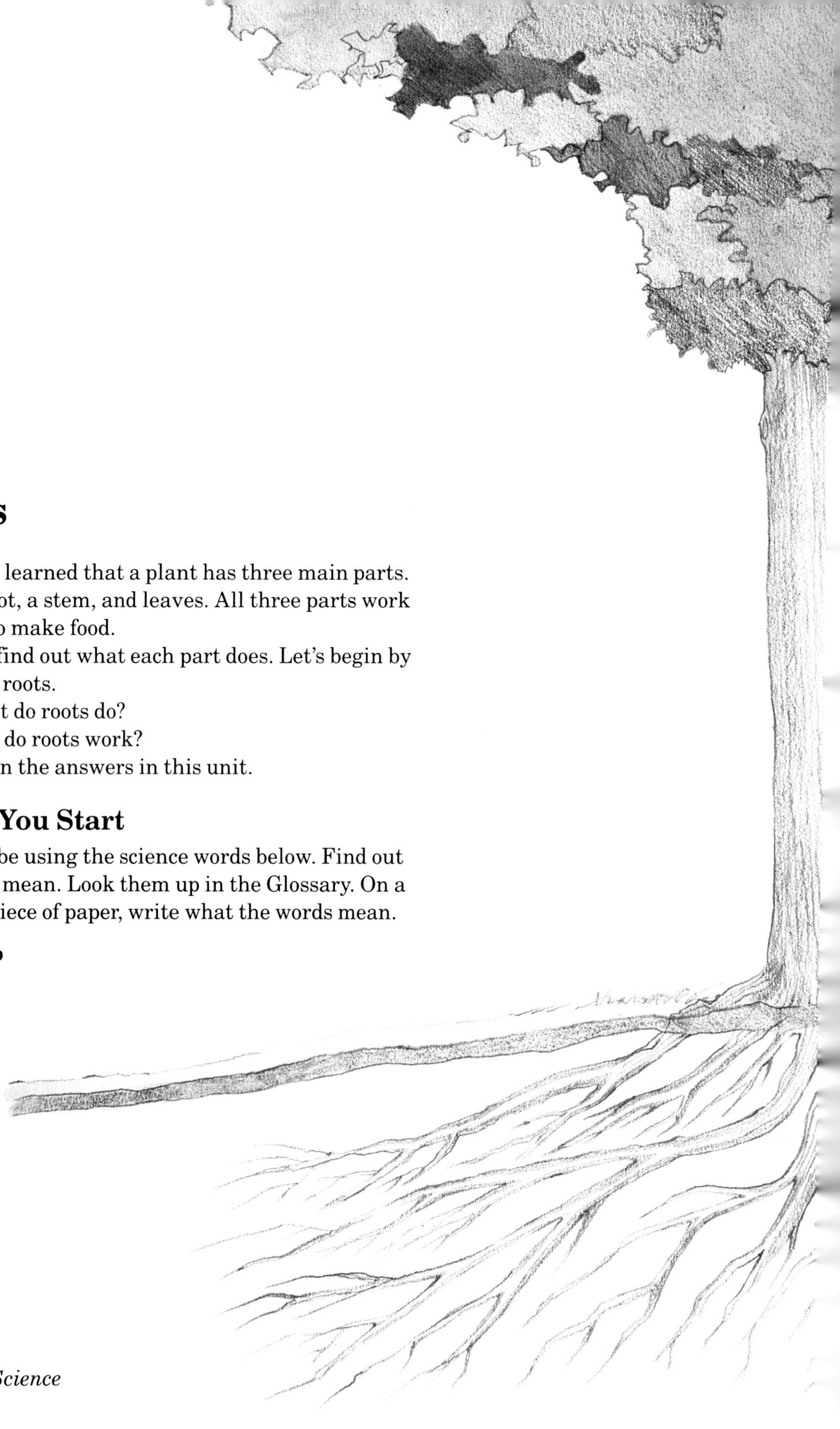

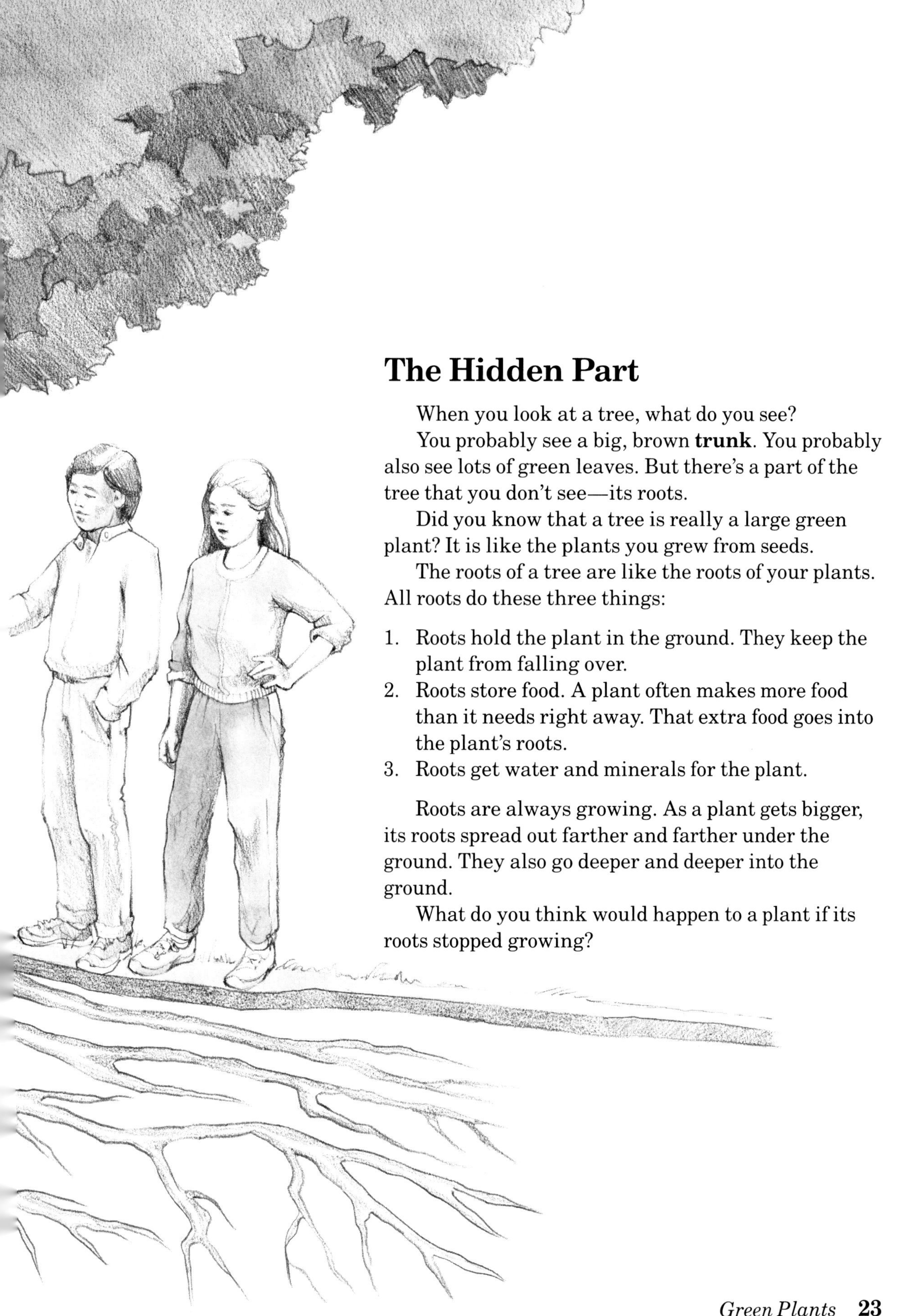

The Hidden Part

When you look at a tree, what do you see?

You probably see a big, brown **trunk**. You probably also see lots of green leaves. But there's a part of the tree that you don't see—its roots.

Did you know that a tree is really a large green plant? It is like the plants you grew from seeds.

The roots of a tree are like the roots of your plants. All roots do these three things:

1. Roots hold the plant in the ground. They keep the plant from falling over.
2. Roots store food. A plant often makes more food than it needs right away. That extra food goes into the plant's roots.
3. Roots get water and minerals for the plant.

Roots are always growing. As a plant gets bigger, its roots spread out farther and farther under the ground. They also go deeper and deeper into the ground.

What do you think would happen to a plant if its roots stopped growing?

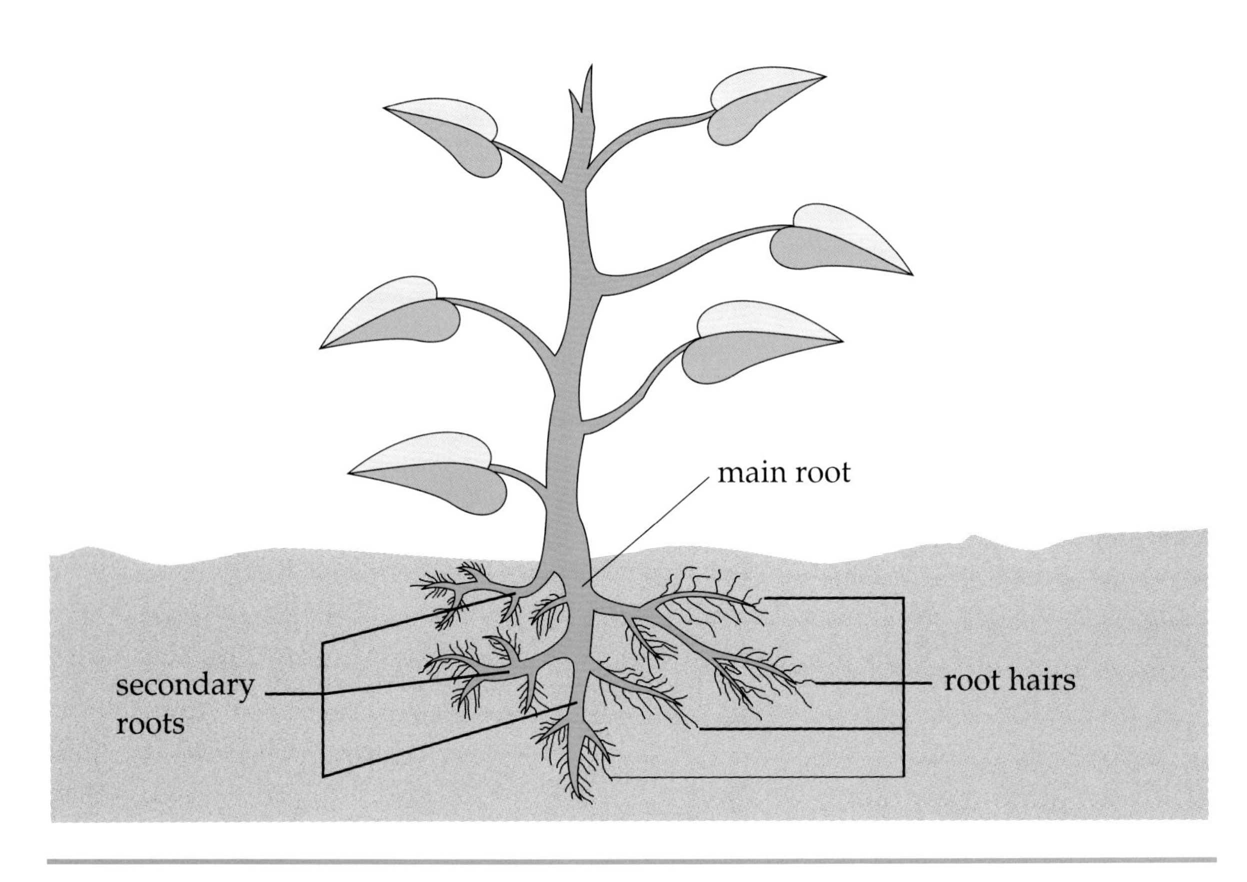

The roots of a seedling

Water Finders

You know that roots get water for a plant. As the plant grows bigger, it needs more and more water. So the roots grow longer, and new roots start to grow.

Think back to when you planted your radish seeds. What was the first thing that came out of each seed?

The first thing was a root. That root was the **main root** of the plant.

As the plant grew bigger, the main root grew longer. It grew down into the soil.

Right after the main root grew, smaller roots started to grow. They grew out of the main root. Those smaller roots were **secondary roots**.

Secondary roots grow in all directions. They get water for the plant.

How does water get into the plant? Here's how: The main root and the secondary roots are covered with tiny **root hairs**. Those root hairs absorb water from the soil. The water moves from the root hairs to the secondary roots. It moves through the secondary roots to the main root. Then the main root takes the water to the stem of the plant.

Look at the picture on this page. It shows the roots of a seedling. What parts absorb water? What parts carry water to the main root? What part takes the water to the stem?

Get into a Root

Did you know that a carrot is a root? It is. And, like all roots, it stores food and gets water.

There are three parts inside a carrot. Water moves up one part. Food moves down another. And food is stored in the third part.

Look at the picture below. It shows the three parts of a carrot.

Now get a carrot. Cut off the top. Then cut the carrot the long way. Look for the three parts.

Find the part of your carrot where food is stored. Find the part where the food moves down. Find the part where water moves up.

On a separate piece of paper, draw the carrot. Show its three parts. Draw an arrow to each part and write what happens in each one.

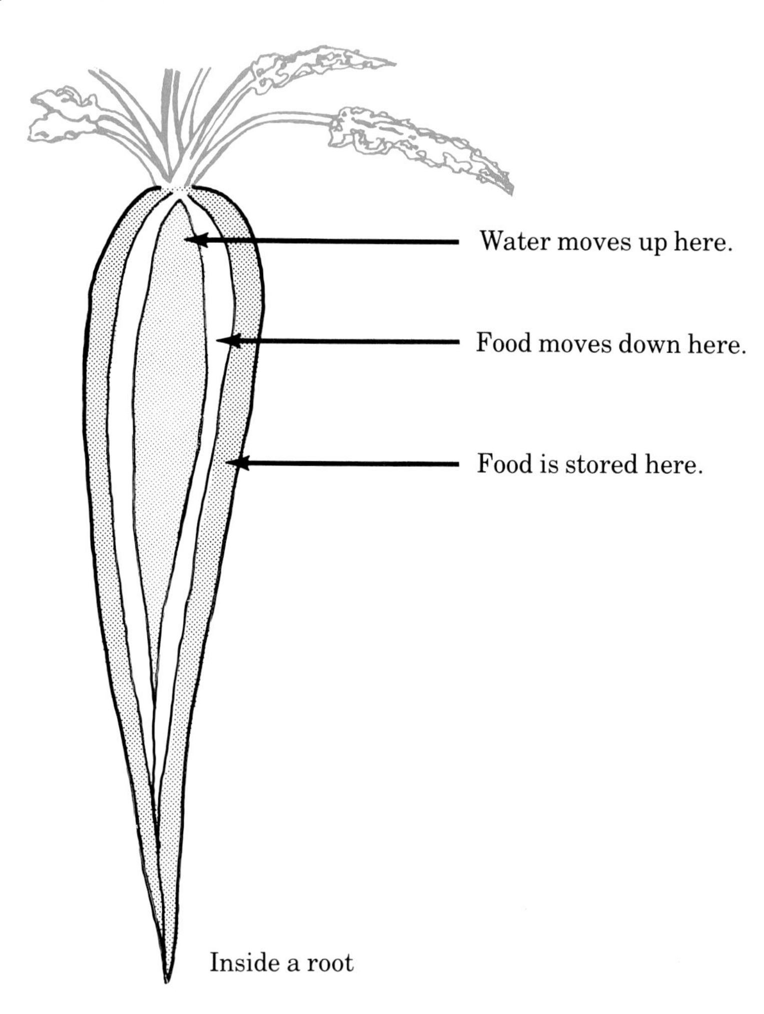
Water moves up here.
Food moves down here.
Food is stored here.
Inside a root

Plant Watch

Choose a bean plant you've grown. It should have at least four leaves.

Carefully take the plant out of its cup. Brush off any soil that sticks to the roots. Straighten out the plant on a piece of paper.

Now look for these parts:

1. Find the first leaves the plant grew. (Hint: Look toward the bottom of the stem.)
2. Find the newest leaves—the ones the plant just grew. (Hint: Look at the tip of the stem.)
3. Find the main root. (Hint: It's the long root in the middle.)
4. Find some secondary roots. (Hint: They grow out of the main root.)

Keep a record of what you find. On a separate piece of paper, draw your bean plant. Be sure to show these parts:

- First leaves
- Newest leaves
- Main root
- Secondary roots

On your drawing, write the name of each part. Your drawing should look something like the picture on this page.

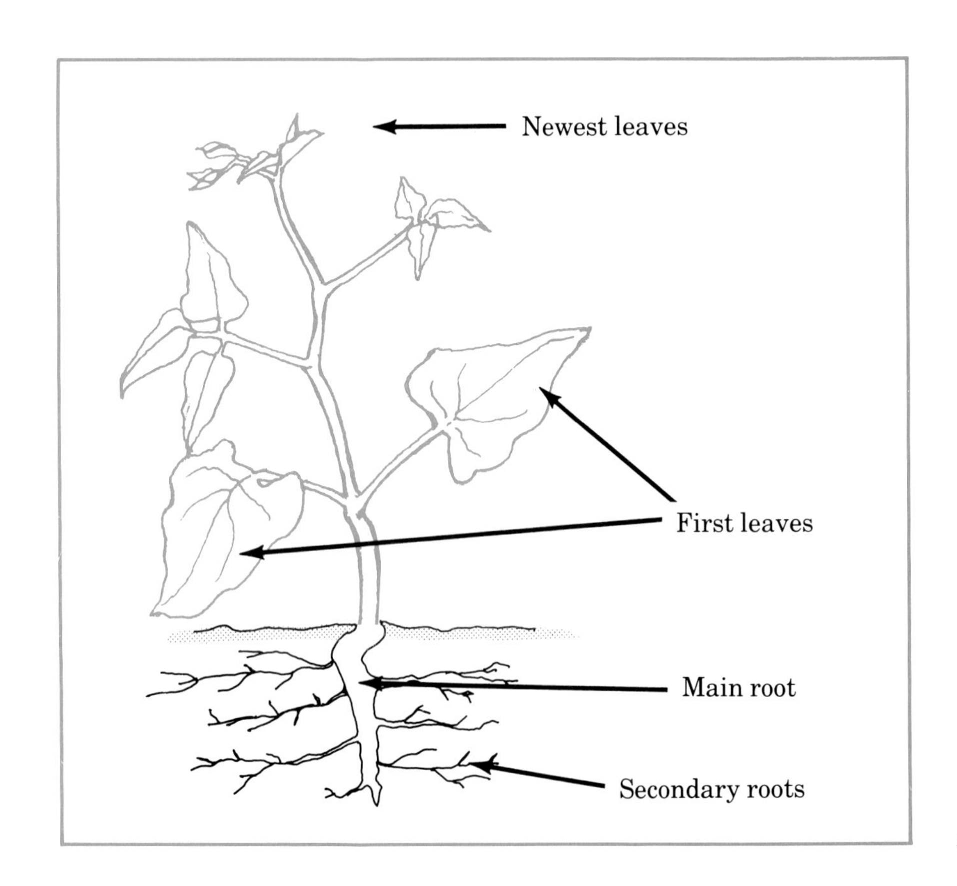

A bean plant

Review

Use what you learned in this unit to answer the questions.

1. What three things do roots do? Pick the three correct answers from the list below:
   a. Hold the plant in the ground
   b. Make food
   c. Store food
   d. Get water and minerals
   e. Get light
2. Show what you know about roots. Use the words below to finish the sentences.

   absorb　　secondary roots
   main root　　stem

The root hairs ________ water. Then water goes into the ________ ________. Next the water goes up the ________ ________ and into the ________.

Check These Out

1. Go to the vegetable section of a grocery store. Find roots you can eat. Some examples are carrots, radishes, and beets. On a separate piece of paper, make a list of the roots you find.
2. Make a carrot cake. Find a recipe in a cookbook, or ask your teacher for a recipe.
3. To watch roots grow, do this: Make a hole in the bottom of a plastic glass. Fill the glass with small colored rocks.

   Get two or three seeds that have just sprouted. Put these seeds inside the glass, next to the sides. Once a day, pour a mixture of water and liquid fertilizer into the glass. (Read the directions on the fertilizer to find out how much to use.)

   As the plants grow, you'll see the roots fill up the glass.
4. Here are more things you may want to find out:
   - What is a tropism? Why do roots grow down?
   - What is osmosis? What does it have to do with roots?

Unit 4

Stems

A bunch of celery, a tree trunk, and a potato: How are these alike? Believe it or not, they are all stems.

A bunch of celery, a tree trunk, and a potato don't look alike at all. But they all do some of the same things. They all do what stems do.

- What do stems do?
- How do stems work?

You'll find out in this unit.

Before You Start

You'll be using the science words below. Find out what they mean. Look them up in the Glossary. On a separate piece of paper, write what the words mean.

1. **pipe**
2. **stalk**

Up and Down the Pipes

You know that plant roots get water. That water goes to the plant stem. Where do you think the water goes from there?

The water goes up the stem and into the leaves. How does it do that?

Think of how water gets into your house. Your house has pipes. Water moves through those pipes. A plant stem also has pipes. The pipes go all the way from the roots to the leaves. Some of those pipes carry water. Some of those pipes carry food.

Look at the picture on this page. It shows how water and food move up and down a stem.

What moves *up* the stem?

What moves *down* the stem?

Right! Water moves up. Food moves down.

A celery stalk is a stem. Like all stems, it has pipes. Where are the pipes in a celery stalk? Find out on the next page.

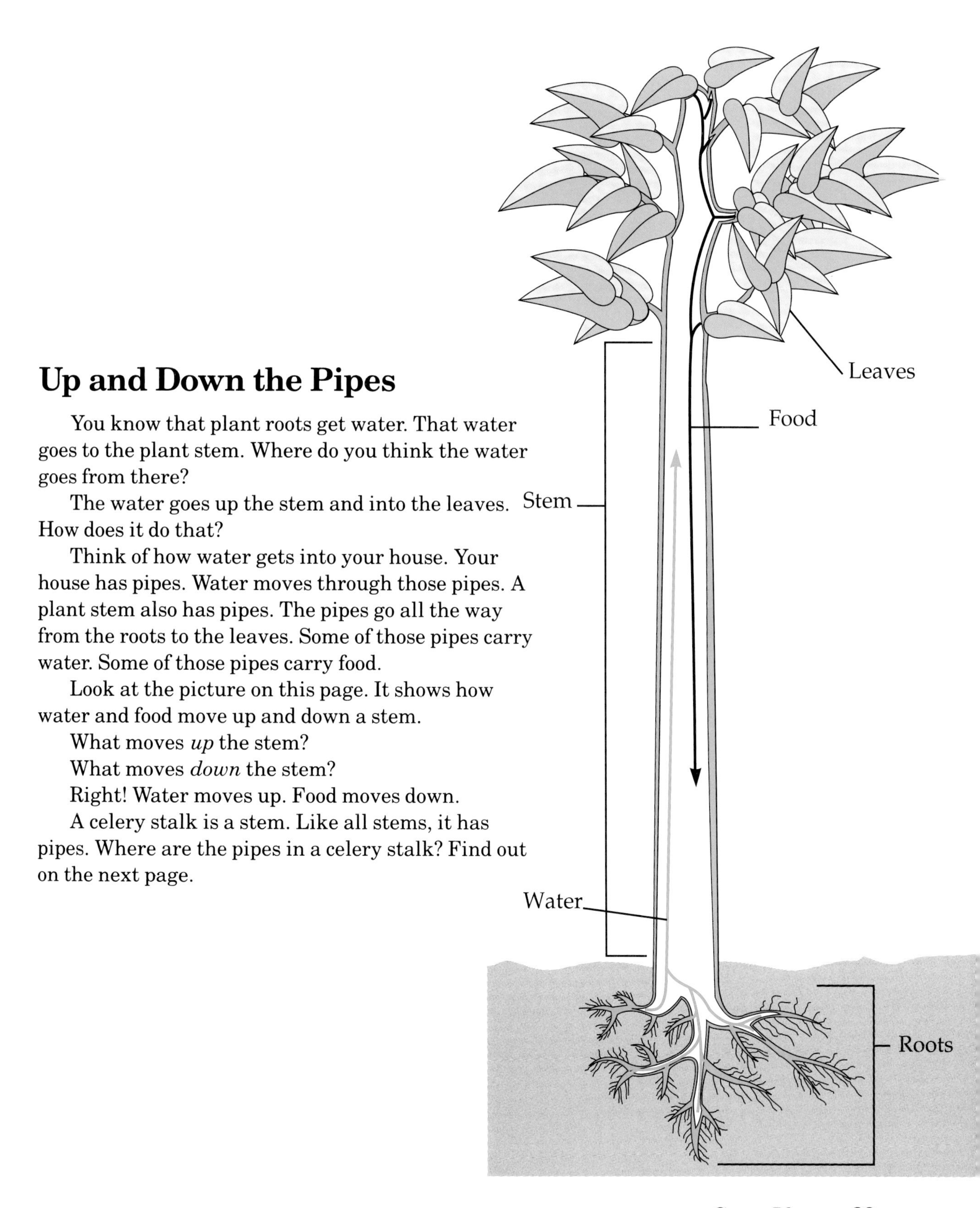

Inside a Stem

You can see how water moves up a plant stem. You can also see some of the pipes. You will need these things:

- Food coloring
- One jar of water
- One celery stalk with leaves
- One knife

1 Dissolve some food coloring in the water.

2 Cut a thin slice off the top of the celery stalk. Don't cut off any leaves.

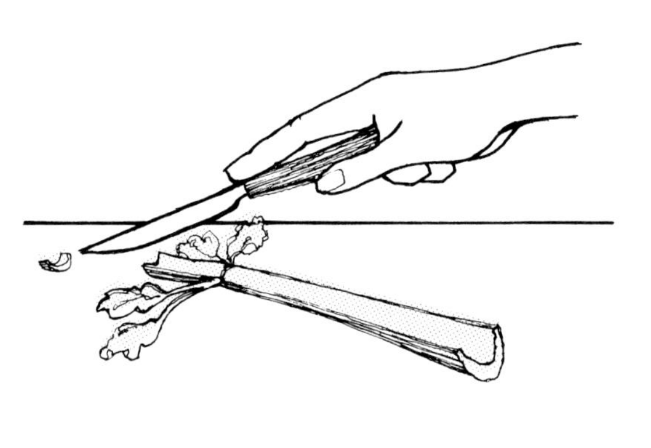

3 Cut off the bottom of the stalk.

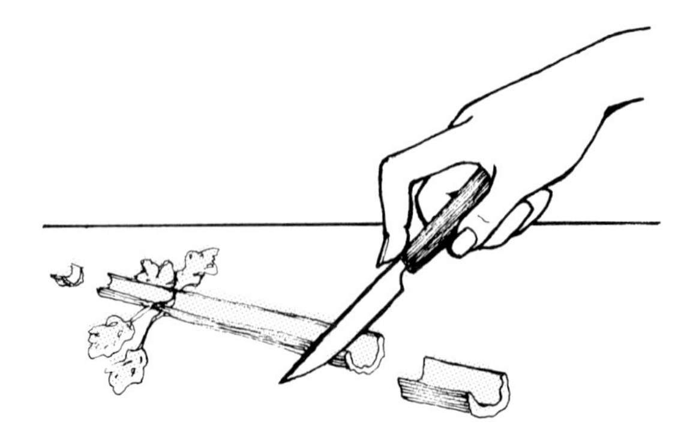

4 Put the celery stalk into the colored water. Then put the jar in a place that gets light. Wait one day. What do you think will happen?

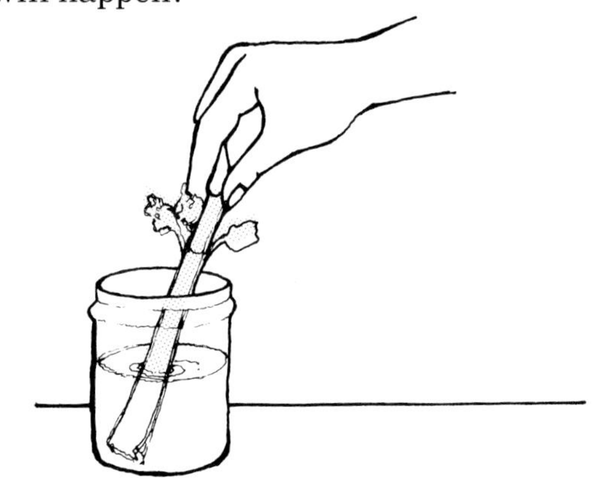

What Happened?

After one day, take the celery stalk out of the water. Then follow these directions:

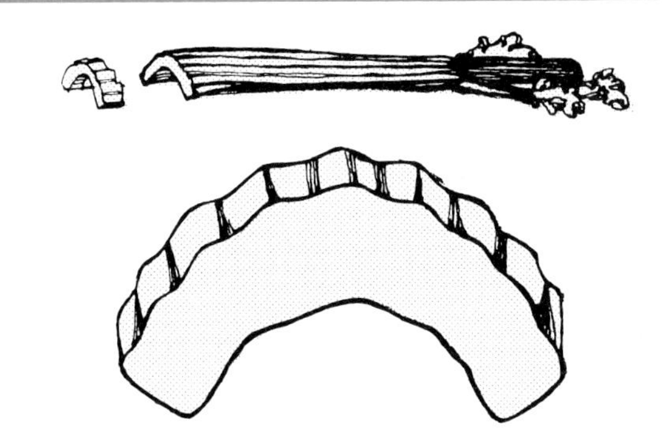

1 Cut a one-inch slice off the bottom of the stalk. Look at the cut end of the stalk. Do you see colored dots? Each dot is the middle of a pipe.

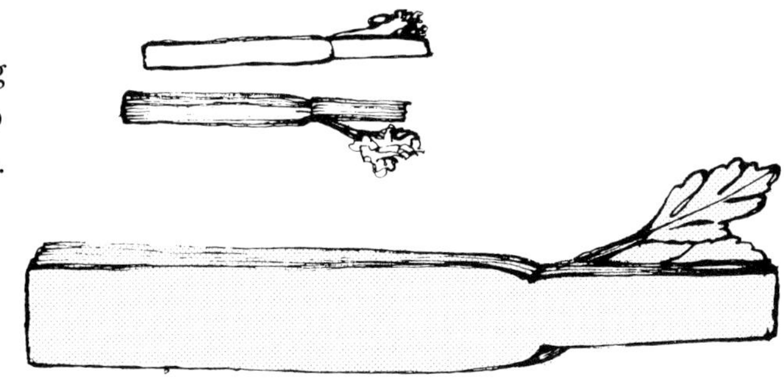

2 Cut the stalks into two pieces the long way. Look at the cut side of one piece. Do you see colored lines? Each line is a pipe.

3 Find a leaf on the stalk that has lots of color. How did the color get to the leaf? Where is the color on the leaf?

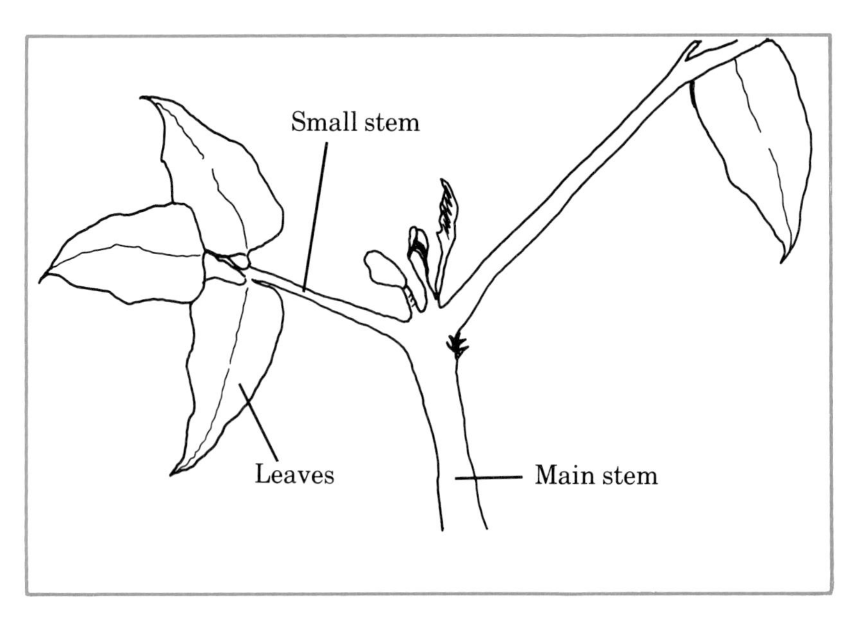

Part of a bean stem

Plant Watch

Get the biggest bean plant you have. It should have one long stem and a few short stems.

Do you see the long stem? It's the main stem. It grows straight up from the soil.

Do you see small stems growing from the main stem? Count those small stems. How many are there?

Do you see leaves growing from small stems? Count those leaves. How many are there?

Keep a record of what you see. On a separate piece of paper, draw your bean plant. Be sure to show these parts:

- Main stem
- Small stems
- Leaves

Also draw anything else you see on the plant.

On your drawing, write the name of each part. Your drawing should look something like the picture on this page.

Review

Use the words you have learned to answer the questions. The words you'll need are listed below.

leaves	roots	stalk
pipes	trunk	stem

1. What parts of the plants get water from the soil?
2. What is one long piece of celery called?
3. What part gets water from the roots?
4. What does the water travel through?
5. What part of a tree is really a large stem?
6. The pipes in a stem go from the roots to what parts?

Check These Out

1. Change the color of a flower. Get a white flower. Cut the stem six inches below the flower. Put the stem in colored water for a day or two. You can also put the stem in different-colored water on the second day. Watch what happens.
2. Look around your house. Find as many things as you can that are made from tree trunks. (A magazine and a wooden table are two examples.) Make a list of what you find.
3. Here are more things you may want to find out:
   - Usually, gravity pulls water down. But water moves up in a stem. Why does it move up?
   - What is xylem? What is phloem? What do they do in a stem?

Unit 5

Leaves

Did you have an apple at lunch today? Or eat some corn chips? Did you have onions in your sandwich? If you did, you ate parts of green plants.

Apples, corn, and onions all contain food that a green plant made. The leaves of the green plant made that food.

- What do leaves use to make food?
- How do leaves make food?
- Why are leaves green?

You'll learn the answers in this unit.

Before You Start

You'll be using the science words below. Find out what they mean. Look them up in the Glossary. On a separate piece of paper, write what they mean.

1. **energy**
2. **gases**
3. **water vapor**

Food Makers

Leaves make the food that a green plant needs. That food keeps the plant alive and growing.

Leaves make food out of two very simple things. One of those things is air. What do you think the other thing is?

Right! The other thing is water.

You learned how a leaf gets water. Water goes into the roots of the plant. From the roots, it moves up the stem and into the leaves.

How does a leaf get air? Suppose you looked at a leaf through a very powerful microscope. You would see tiny holes all over the leaf. Most of the holes are on the underside of the leaf.

Air goes into the leaf through those holes. The air mixes with the water in the leaf. Then the leaf is ready to make food for the plant.

A plant uses the food that its leaves make. People and other animals use that food too.

Think of the food you eat. Do you eat fruits and vegetables? Seeds and grains? Those are all parts of green plants. They grow because of the food that leaves make.

Do you eat beef? Pork? Chicken? Those foods come from animals. Those animals grow because they eat the food that green leaves make.

So green plants give us the food we eat. What do you think the world would be like if we didn't have green plants?

In and Out

To make food, green leaves take in two things—water and air. But they also give out two things. Those two things come out of the tiny holes in the leaves.

Air is made up of different gases. One of those gases is **carbon dioxide**. When leaves take in air, they use carbon dioxide to make food. As they make food, they also make another gas that is part of air. That gas is oxygen. The leaves also put oxygen into the air.

So one of the things that leaves give out is oxygen. People and other animals need oxygen. Without it, they couldn't live.

Look at the picture on this page. Answer the questions above the picture. Use the words *oxygen* and *carbon dioxide*.

What do leaves take in?
What do leaves give out?

Experiment 3

What else do leaves give out?

Materials

- One tomato plant that has lots of leaves
- Water
- One plastic bag big enough to fit over the plant
- One piece of string or a plastic-bag tie
- A sunny window

Procedure

1. Water the plant. Then put the plastic bag over the plant. Make sure all the leaves are covered.
2. Tie the bag around the stem.
3. Put the plant next to a sunny window. Leave it there for a day.

Observations

1. After a day, what do you see on the inside of the plastic bag?
2. Where do you think it came from?

Conclusions

Leaves give out oxygen. What else do they give out?

What Goes Up Must Come Out

In Experiment 3 on page 37, you found out that plant leaves give out water. Where does the plant get that water?

Right! The plant gets that water from the soil. Suppose you water a plant. Water goes into the soil. The roots absorb water. The water moves up the stem and into the leaves.

The leaves use some of the water to make food. What do you think happens to the rest of the water?

The rest of the water goes out of the leaves.

You usually can't see the water that comes out of the leaves. That's because the water is in the form of water vapor. Water vapor is a gas. (In Experiment 3, you saw the water because of the plastic bag. The bag trapped the water vapor. And the vapor turned back into water.)

Look at the picture of a plant on this page. It shows how water moves through a plant and out of the leaves. There are four steps. What are they?

4. The rest of the water goes out the leaves.
3. The water goes into the leaves. The leaves use some of the water to make food.
2. The water moves up the stem to the leaves.
1. The roots absorb the water from the soil.

How water moves through a plant

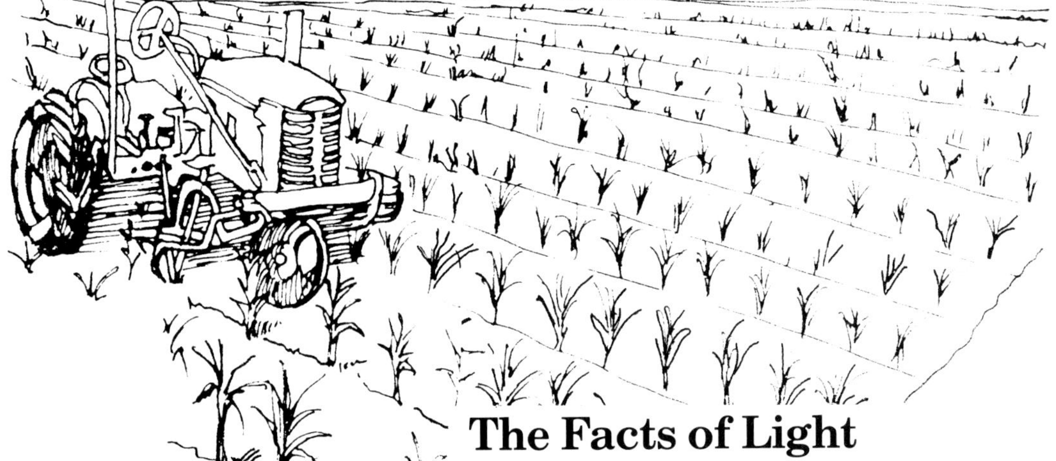

The Facts of Light

Think of how you make a cake. You mix together some dry things and some wet things. You put the mixture into a hot oven. The heat energy in the oven turns the mixture into a cake.

Now think of how leaves make food. Leaves mix water with air. But they can't turn the water and air into food without energy. Where do you think leaves get energy?

Leaves get energy from light. That light usually comes from the sun.

There's a special name for the way green leaves make food. That name is **photosynthesis**. *Photo* means "in light." (A *photograph* is a picture taken in light.) *Synthesis* means "making something." What do you think *photosynthesis* means?

Photosynthesis happens in green plants because of something that's in their leaves. That something is **chlorophyll**. (Many plants have chlorophyll in their stems too.) Chlorophyll traps light energy. Then leaves use that light energy to turn water and air into food.

A plant that doesn't get any light looks different from a plant that gets light. That's because light does something to chlorophyll.

What does light do? You'll find out in the next experiment.

Experiment 4

What does light do to chlorophyll?

Materials

Four lima beans

Soil

One Styrofoam cup

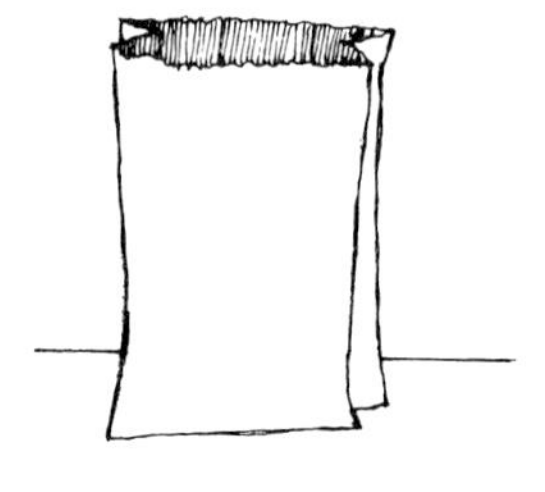

One paper bag

Procedure

1. Soak the beans overnight in water.

2. Turn the cup over. With a pencil, poke three holes near the bottom of the cup.

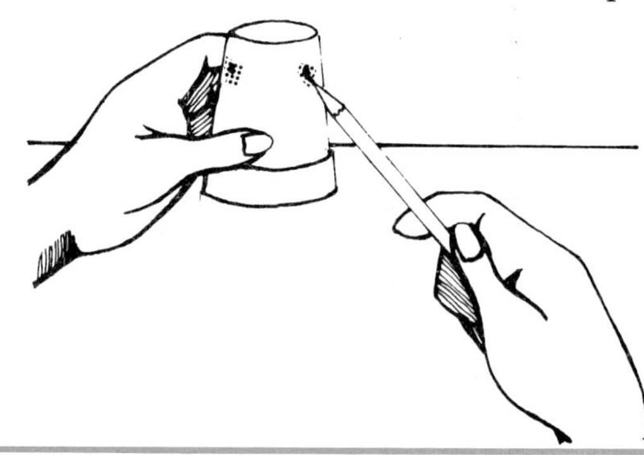

3. Fill the cup with soil. Put the soaked seeds on the soil. Cover the seeds with a thin layer of soil. Water the soil.

4. Put the cup into the paper bag. Make sure the cup stands up in the bag.

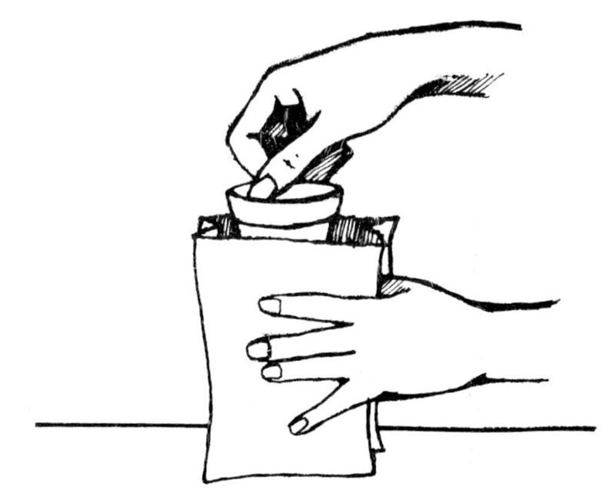

5. Close the bag so that no light can get into it. Put the bag in a place away from light, such as a cupboard. Leave the bag there until the seeds sprout and grow leaves (about four or five days).

6. When the seedlings have leaves, take the cup out of the bag. Put the cup in a sunny window. (Or put it under strong fluorescent light.) Leave the seedlings there for at least an hour.

Observations

1. What color were the seedlings when you took them out of the bag?
2. What color are the seedlings after they get light?

Conclusions

What does light do to chlorophyll?

Plant Watch

Choose a plant you've grown. It should have lots of leaves. Notice that the leaves are spread out. And most of the leaves face one way. They face the light. That way, each leaf gets as much light as it can.

Turn the plant so that the leaves are facing you. On a separate piece of paper, keep a record of what you see. Answer these questions about the plant.

1. How many leaves are on the plant?
2. Are some leaves bigger than others?
3. Are some a darker color than others?
4. Why do you think that is?
5. Does any leaf block another leaf's sun?
6. If so, which leaf is a darker color?

Draw a picture of your plant.

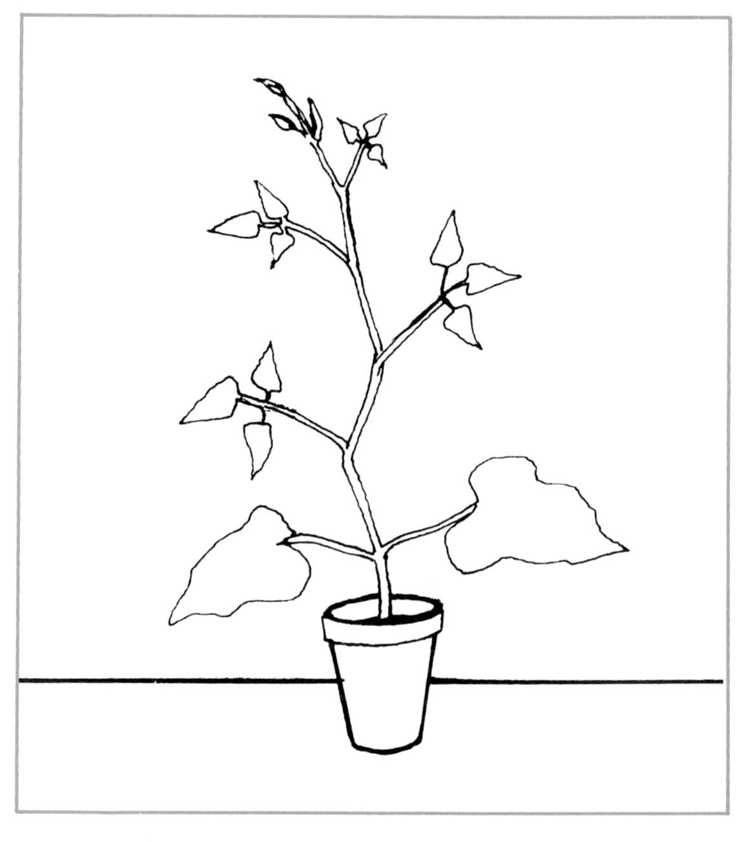

Leaves spread out and face the light.

Review

How much do you know about photosynthesis? Match the words on the left with the clues on the right.

1. chlorophyll	a. What green leaves give out
2. oxygen and water vapor	b. Photosynthesis can't happen without this.
3. green	c. This moves from the roots to the leaves.
4. water	d. What traps the energy in light
5. light	e. The color chlorophyll turns
6. air	f. Green leaves get this from light.
7. energy	g. What goes into a leaf through tiny holes

Check These Out

1. Make a leaf rubbing. Cover a leaf with a very thin piece of paper. Lightly color the part of the paper that's over the leaf. Use crayon, chalk, or pencil. You'll get a picture of the leaf. On it, write the name of the plant the leaf came from.
2. Make pressed leaves. Get some leaves. Put each leaf inside a folded paper towel. Then put the towels inside a thick book. Keep them there for a few days. When the leaves are flat and dry, tape them to heavy paper. Under each leaf, write the name of the plant it came from.
3. Here are more things you may want to find out:
   - Why do leaves change color in the fall?
   - Some plants, such as mushrooms, don't have chlorophyll. How do those plants get food?
   - Plants need carbon dioxide. People need oxygen. How do they get those things from each other?
   - How do plants reproduce?

Unit 6

New Plants from Old Plants

Start Plants from Roots

You have started plants from seeds. There's another way to start plants. You can start new plants from parts of old plants.

You can start some kinds of plants from roots. You will need these materials:

- One carrot with the stems and leaves cut off
- One knife
- One dish
- Water

1 Cut a one-inch piece off the top of the carrot.

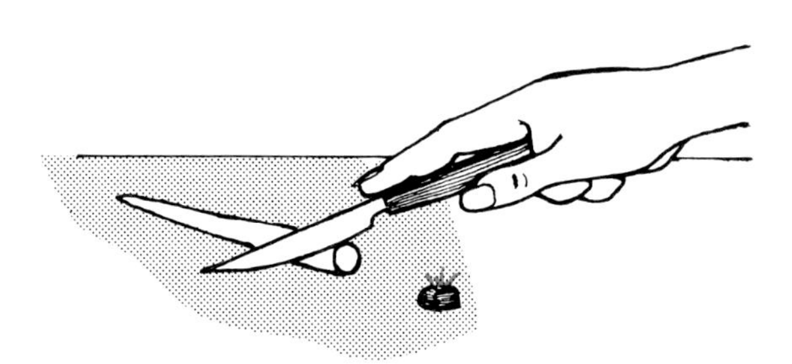

2 Put the carrot top into a dish. Cover most of the carrot top with water.

3 Put the dish in a place that has light. Leave it there. If the water starts to dry up, add more water. In a few days, roots and leaves will grow out of the carrot top.

What other plants can you start from roots? Find out. Put the tops of roots from three different plants in water. What plants did you choose? Which ones grew?

Start Plants from Stems

You can start some kinds of plants from stems. Here's one way: Cut off a stem from a plant. The stem should have some leaves on it. Put the stem in a glass of water. In a few days, roots will grow out of the stem.

These pictures show another way to start a plant from a stem. You will need these things:

- One bean plant that you grew
- Scissors
- One Styrofoam cup
- Soil
- Water

1 Cut a stem off the bean plant.

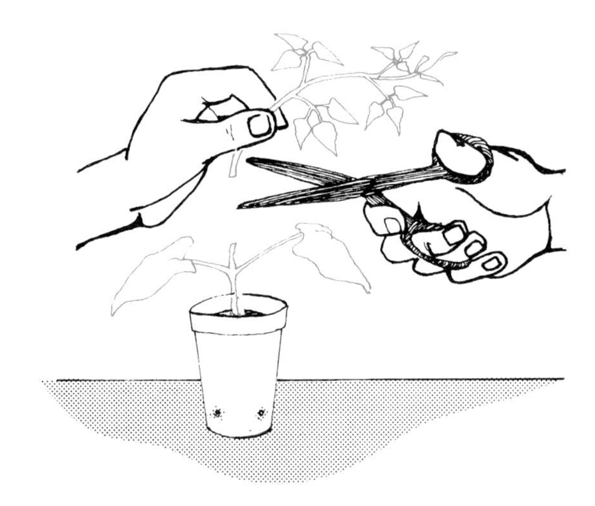

2 Poke holes near the bottom of a paper cup and fill the cup with soil. Wet the soil.

With a pencil, make a hole in the wet soil. Put the stem into the hole. Carefully pack the soil around the stem so the stem stands straight up.

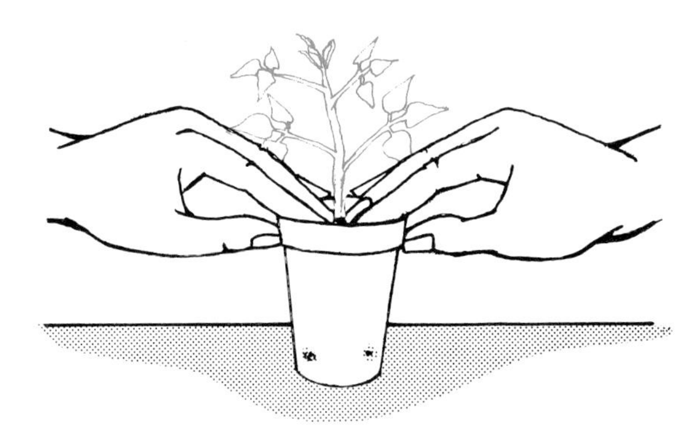

3 Put the cup where there is light. In a few days, roots will grow out of the stem under the soil.

What other plants can you start from stems? Put stems from three different plants in water. What plants did you use? Which ones grew?

Start Plants from Leaves

You can start some kinds of plants from leaves. Try it with one of these plants: jade plant, ivy, or philodendron. You will need these materials:

- One leafy plant
- Scissors
- One Styrofoam cup
- Soil
- Water

1 Choose a healthy leaf. Cut off the leaf and its stem.

2 Poke holes near the bottom of the paper cup and fill the cup with soil. Wet the soil.

3 Put the leaf stem into the wet soil. Carefully pack the soil around the stem so the leaf stands straight up.

4 Put the cup where there is light. Water the soil when it gets dry. In a few days, roots will grow out of the stem under the soil.

What other plants can you start from leaves? Put leaves from three different plants into soil. What leaves did you use? Which ones grew?

Plant Watch

Choose a plant that you started from a plant part. It can be a plant grown from a root, a stem, or a leaf. On a separate piece of paper, keep a record of facts about your plant.

1. What is the name of the plant?
2. What part of the plant did it come from: root, stem, or leaf?
3. When did you plant it?
4. What was the *first* thing you saw that let you know your plant was growing? Could you see roots? Did the stem get longer? Did a new leaf sprout?
5. Of the plants that you grew from plant parts, which plant grew the fastest?
6. Draw a picture of the plant.

Show What You Learned

What's the Answer?

Choose the correct answer for each question. There may be more than one correct answer to a question.

1. What does a seed need to sprout?
   a. Water
   b. Warmth
   c. Chlorophyll
2. What are the three main parts of a plant?
   a. Leaves, root, and stem
   b. Root, seeds, and leaves
   c. Leaves, stalk, and stem
3. What do roots do?
   a. Hold a plant in the ground
   b. Make food
   c. Get water
4. How does water move up and down a stem?
   a. Through nutrients
   b. Through root hairs
   c. Through pipes
5. What do leaves use to make food?
   a. Seeds, oxygen, and other gases
   b. Water, carbon dioxide, chlorophyll, and light energy
   c. Soil, darkness, chlorophyll, and a cup

What's the Word?

Give the correct word for each meaning.

1. A very young plant
   S ________
2. Break down into very small pieces in water
   D ________
3. Something plants need to stay healthy
   N ________
4. Soak up water
   A ________
5. A stem on a celery plant
   S ________
6. The way that green plants make food
   P ________
7. The power that makes things change
   E ________

Congratulations!

You've learned a lot about green plants. You've learned

- How to start plants from seeds or other plants
- What plants need to grow
- How plants make food
- And many other important facts about plants

ANIMALS

What kind of living things are animals? How are they like other living things? How do animals stay alive? In this section, you'll learn many facts about these living things. And you'll learn how animals play an important part in our lives.

Contents

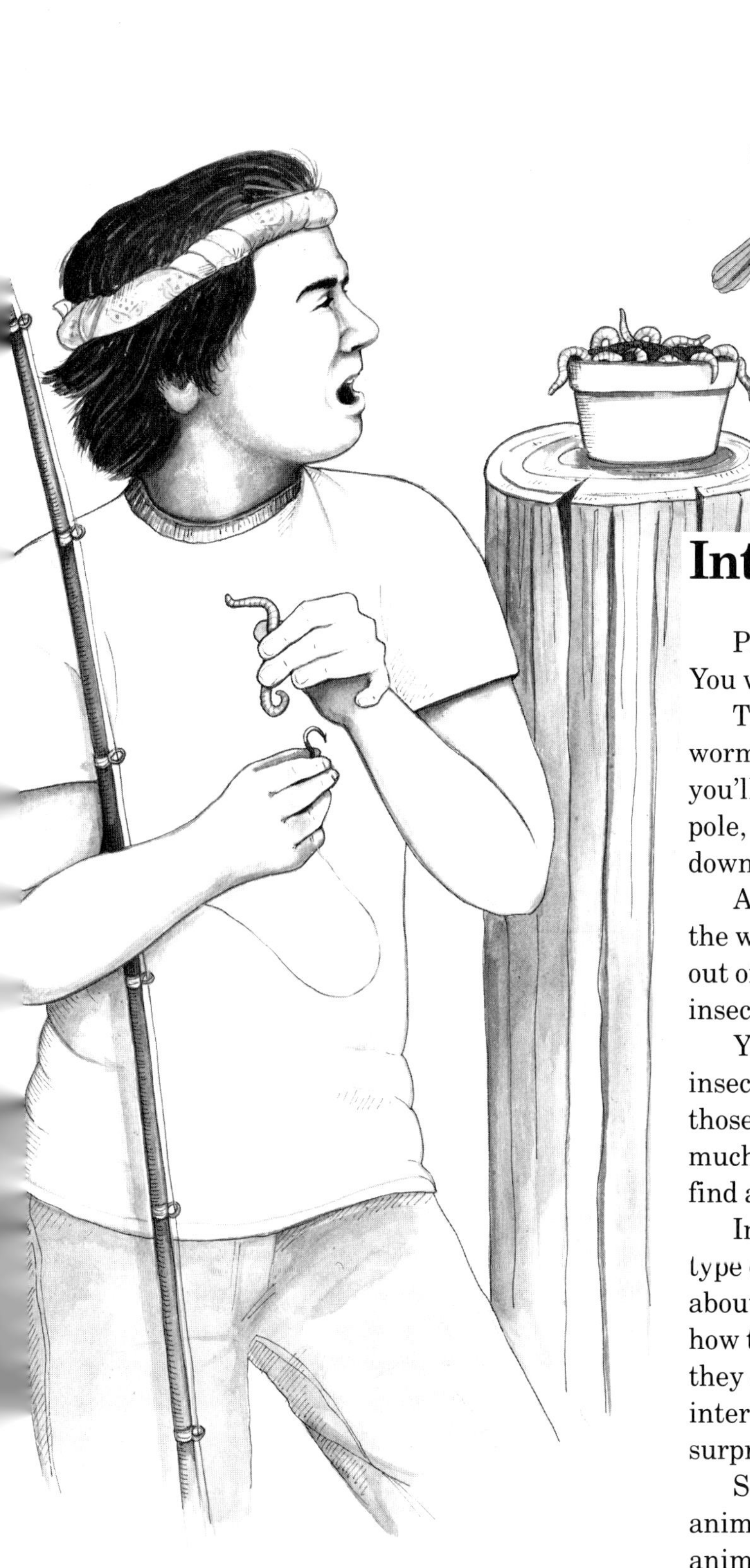

Introduction

Picture this: You decide to go fishing. You will eat the fish you catch for dinner.

The fish eat worms, so you get some worms for bait. Next you get the things you'll need to catch your fish—fishing pole, hooks, net, and so on. Then you go down to the river.

At the river, you see fish swimming in the water. Some of them suddenly jump out of the water. They catch and eat insects that fly above the water.

You, the worms, the fish, and the insects are all animals. You don't look like those other animals, but you are very much like them. For one thing, you must find and eat food to stay alive.

In this section, you'll learn about a type of living thing: animals. You'll learn about many different animals. You'll learn how they are alike and different, and what they do to stay alive. You'll also learn interesting science facts that will surprise you.

Someday you may want to raise animals. You may need to get rid of animal pests. You may want to watch wild animals to learn about them. You'll be able to do those things if you understand what certain animals need and how they act.

This section can help you find out.

Unit 1

Living Things Called Animals

Earth is different from other planets in one very important way: Many, many kinds of things can live on it. Some of those things we call *plants*. Some we call *animals*.

Animals live almost everywhere on Earth. Some even live inside other animals. No one knows how many different kinds of animals there are. So far, scientists have studied about a million different kinds. But new kinds are always being found.

- When is a living thing called an animal?
- How do scientists describe an animal?

You'll find out in this unit.

Before You Start

You'll be using the science words below. Find out what they mean. Look them up in the Glossary that's at the back of this book. On a separate piece of paper, write what the words mean.

1. **classify**
2. **traits**

What Is an Animal?

Millions and millions of different kinds of living things are on Earth. Scientists divide those living things into groups. What are the two main groups?

Right! The two main groups are *plants* and *animals*.

How do scientists decide when a living thing is an animal? They see if that living thing has the *traits* of an animal. What traits do you think animals have?

All animals have these traits:

- Animals can move themselves. Most of them can also move from place to place.
- Animals must eat food. They must also get water and oxygen or they will die.

Many kinds of animals are very different from each other. Their bodies don't look alike. And they don't act in the same ways. But many other kinds of animals are very much alike. Their bodies look different, but they have certain traits that are the same.

Look at the human and the birds in the picture. They each have two legs. That is a trait both animals have. How else is the human's body like the bird's body?

Animal Traits

Think of a cat. Now think of a dog. They are different kinds of animals. Yet they have traits that are the same. What is one trait that both cats and dogs have?

Cats and dogs also have traits that are different. What is one trait that cats have that dogs don't?

Scientists do just what you did when they study an animal. They study its traits. They see which traits are like other animals'. And they look for traits that belong to it alone. They say an animal is a certain kind (such as a cat) when it has traits that no other kind has.

Scientists *classify* animals by their traits. They say that animals with the same traits make up a group. For example, animals with skeletons inside their bodies make up one group. Animals without such skeletons make up another group.

The pictures on these pages show some kinds of animals. How might you group them? Think of some traits those animals have. Write those traits on a separate piece of paper. Under each trait, write all the animals in the pictures that have that trait. Some animals will be in more than one group. Suppose you write the trait "Lives in water." You would list both "clam" and "fish" under that trait.

bird

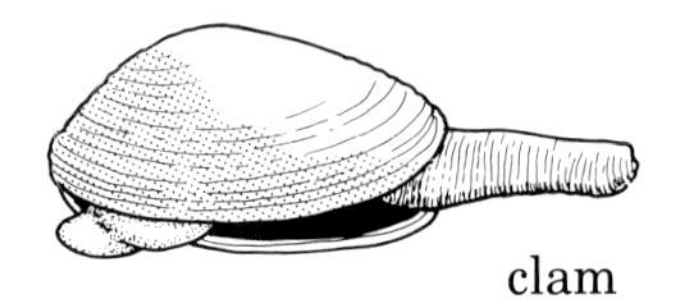

clam

butterfly

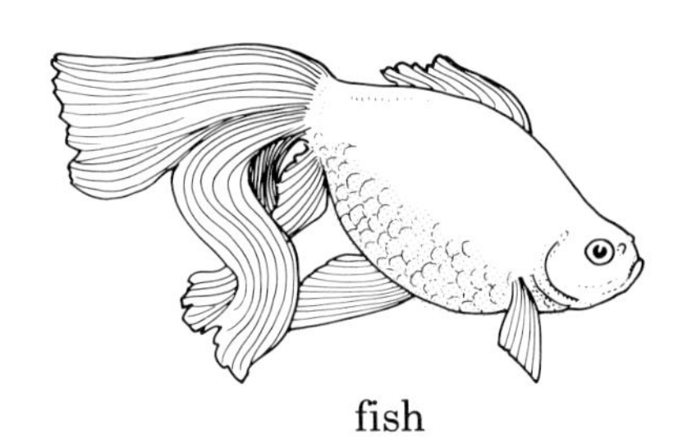

fish

gorilla

humans

dog

What Kind of Animal Are You?

Now think about yourself. You are a *human*. What traits do humans have that other animals have?

Think of traits that describe your body. For example: *Has an inside skeleton*. Think of traits that describe the way you act. For example: *Lives in homes that are built*. Then think of other animals that have those traits too. (You can choose animals that are not in the pictures.)

On a separate piece of paper, write each trait. Then write some animals that also have that trait.

Think About It

When scientists study a kind of animal, they give it a **scientific name**. All scientists then use that name when they talk about that animal. Do you know the scientific name for humans?

The scientific name for humans is *Homo sapiens*. The words are Latin. *Homo* means "human" and *sapiens* means "intelligent."

Here are some scientific names for other animals. Can you guess the animal for each one? (The answers are upside down.)

1. *Canis familiaris*
2. *Felis catus*
3. *Equus caballus*

Answers

1. dog
2. cat
3. horse

Review

Use what you learned in this unit to answer the questions. Then check your answers. The page where you can find the answer is listed after each question.

1. Scientists divide living things into two groups. What are those groups? (page 53)
2. What do you call the certain ways an animal looks or acts? (page 52)
3. What are the traits that all animals have? (page 53)
4. What is it called when scientists group animals by traits? (page 54)
5. What is the scientific name for humans? (page 55)

Check These Out

1. Make a Science Notebook for this section. Use it to keep a record of what you learn about animals. Put your list of glossary words and their meanings in the notebook. Also keep your notes from the Animal Watch sections in it. You can put anything else you learn about animals in your notebook too.
2. What are the traits of living things? Find out. Then give a talk about the differences among a rock, a plant, and an animal.
3. Make a poster that shows the groups humans can be classified in. Paste pictures of other animals that also belong in those groups.
4. Animals are one *kingdom* of living things. Scientists say there are two or three more. Find out what they are.
5. Make your own *system of classification* and classify the people in your building or school.
6. Make a list of traits that belong just to you.
7. As you work your way through this section, you may want to find out more about animals. You can find out more by looking in an encyclopedia or by getting books about animals in the library. You can also talk to an expert, such as a zoologist, naturalist, or veterinarian.

   Here are some things you may want to find out:

   - Who was Linnaeus? What is his system of classification?
   - What are the smallest living things? What do they look like?
   - How do scientists name animals?
   - Who are the Leakeys (Louis, Mary, and Robert)? What did they discover?

Unit 2

A Look at the Animal Kingdom

You learned that scientists study an animal by describing its traits. Then they classify that animal into several groups. One of those groups is called a *phylum*.

Many different kinds of animals are in each phylum. But all those animals have some important traits that are the same. Scientists divide all the animals on Earth into about 25 *phyla*. (*Phyla* means more than one phylum.) Each phylum has its own special traits.

- What are the main phyla of animals?
- What are some animals in the main phyla?
- What are some traits of the main phyla?

You'll find out in this unit.

Before You Start

You'll be using the science words below. Find out what they mean. Look them up in the Glossary. On a separate piece of paper, write what the words mean.

1. **body cavity**
2. **class**

Sponges don't move from place to place. They attach themselves to rocks and other things.

sponges

Sponges

Sponges belong to the phylum *porifera*. Sponges don't move around like many animals. They live attached to rocks and other things.

The sponge is a *simple* animal. That means it has few body parts. Its body is like an empty sack. The inside of the sack is its **body cavity**.

Look at the diagram of the sponge's body. See how its body is like a sack? What else do you notice about the sponge's body?

It's full of *pores* (holes). That's why the sponge phylum is called porifera. *Porifera* means "having holes." Pores are a trait of the phylum.

A sponge doesn't have a mouth or teeth. How do you think it gets food?

Here's how: Sea water is full of tiny animals. Those animals are food for the sponge. The sponge pulls the water into its body. It traps the animals inside its body. Then it pushes the water out.

Look at the diagram of the sponge again. It shows that water goes in through the pores. The large arrow shows where the water goes in and where it goes out. What does the water go out through?

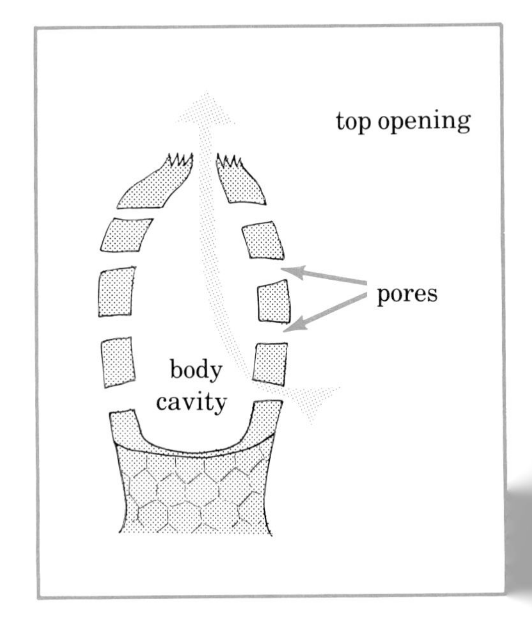

The body of a sponge

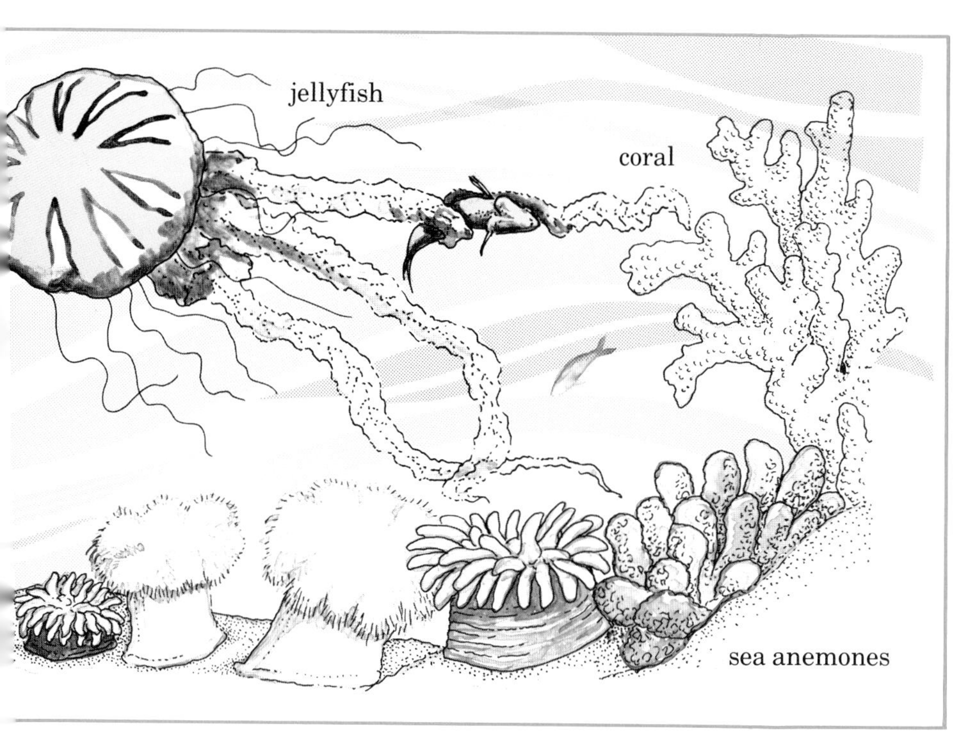

The jellyfish, coral, and sea anemone are three kinds of coelenterates. Which of those animals can move from place to place?

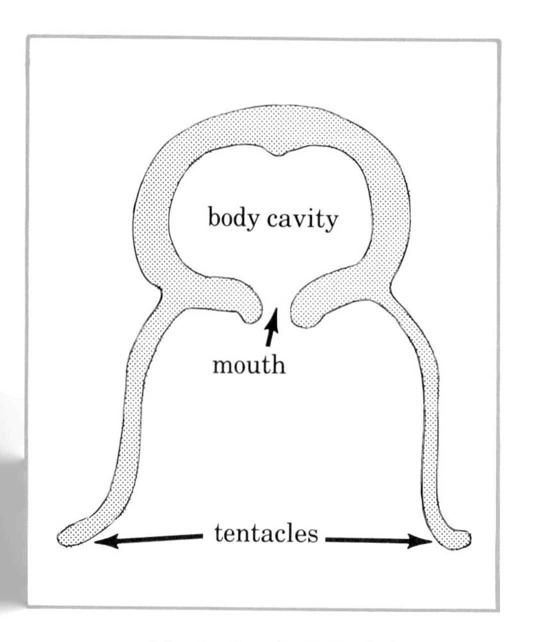

The body of a jellyfish

Coelenterates

Jellyfish, sea anemones, and corals look very different on the outside. But inside, they are very much alike. That's why they belong in the same phylum: *coelenterate*. Jellyfish, sea anemones, and corals belong to different *classes*. A class is a group of animals in the same phylum.

Some people call coelenterates *stinging cell animals*. That's because they have **tentacles** that can sting. Those tentacles are used like arms.

The coelenterate uses its tentacles to catch food. It spreads out its tentacles in the water. It stings and stuns the fish that swim into its tentacles. Then it wraps its tentacles around the fish and pulls the fish into its mouth.

Look at the diagram of a jellyfish's body. How is the jellyfish's body like the sponge's body?

How is it different from the sponge's body?

Now look at the pictures at the top of these pages. How else is the jellyfish different from the sponge?

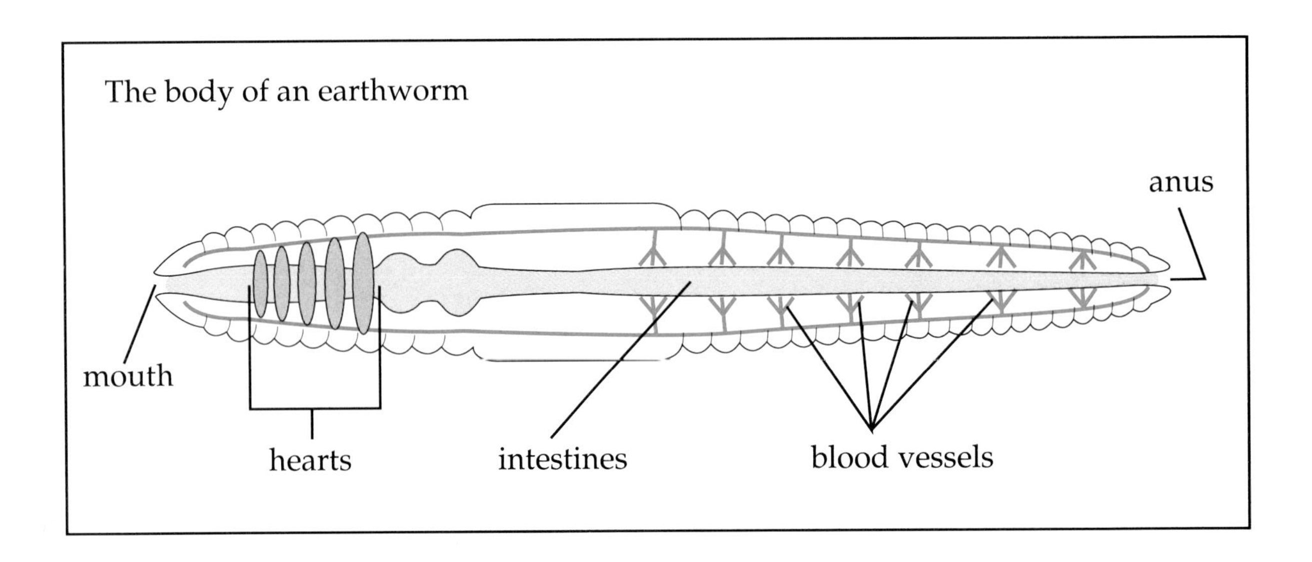

Worms

You learned that animals with simple bodies have very few parts. Now you'll learn about animals whose bodies are not simple. Worms are some of those animals.

Flatworms, roundworms, and *segmented worms* are three phyla of worms. Flatworms have long flat bodies. Most flatworms are *parasites* that live inside the bodies of other animals. Most roundworms are also parasites. What do you think a roundworm's body is shaped like?

That's right! Roundworms have bodies that are long and round.

Segmented worms have bodies that are made up of **segments** (small sections). They also have more kinds of body parts than the animals you have studied so far. Earthworms are segmented worms.

Look at the diagram of the earthworm's body. What parts does the earthworm have that the jellyfish doesn't?

The earthworm has special parts for handling food. The *mouth* takes food into the body; the *intestines* digest food; and the *anus* pushes out wastes. Those parts make up the earthworm's **digestive system**.

The earthworm also has two kinds of parts that move blood through its body. Those parts make up its **circulatory system**. Look at the diagram. What are those parts?

Right! Those parts are the hearts and the blood vessels.

Mollusks

This next group of animals belongs to the *mollusk* phylum. *Mollusk* means "soft body." Clams, octopuses, squid, snails, and slugs are mollusks. They all have these traits: soft body, big foot or tentacles, digestive and circulatory systems.

Mollusks live in many different places. Some live in fresh water or salt water. Some live in mud or sand. And some live on dry land. Look at the pictures of the mollusks. Which mollusk lives on land?

Some kinds of snails live on land. Many kinds of snails live in water. Octopuses and clams live in water.

Most mollusks have hard shells that protect their soft bodies. Mollusks such as clams have two shells.

But some mollusks don't have an outside shell at all. The octopus is an example of such a mollusk.

What other mollusks do not have outside shells? (The answer is upside down.)

octopus

clam

snail

Answer

Squid and slugs are some other mollusks that don't have outside shells.

The body of a squid

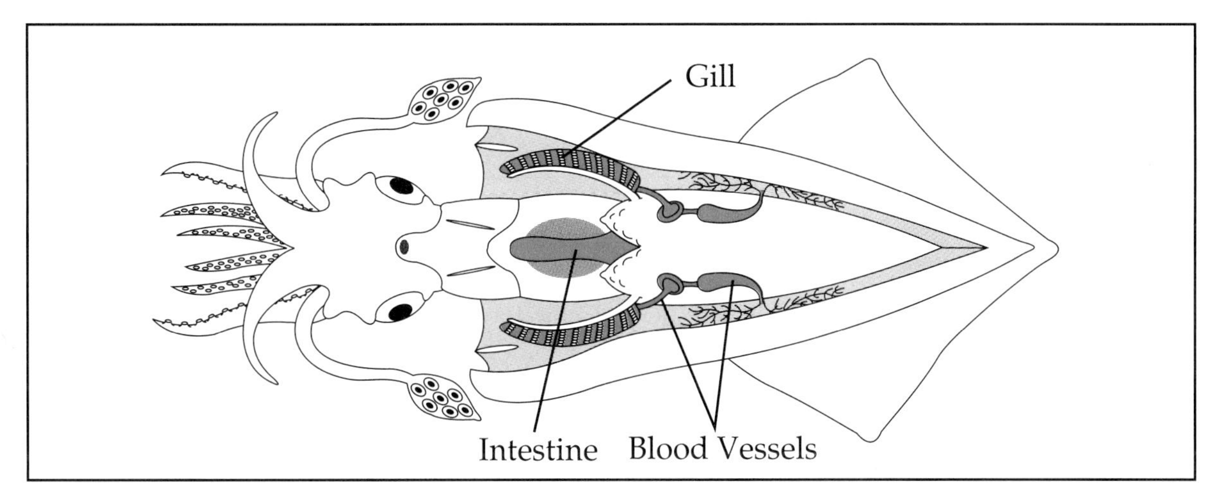

Arthropods

A spider, a bee, and a crayfish—how are they alike? Look closely at the pictures on this page. What body parts do all those animals have that mollusks and worms don't have?

Right! They have *legs*.

The animals in the pictures belong to the largest animal phylum—*arthropod*. *Arthropod* means "jointed legs." Those animals have legs that are made up of several parts.

Suppose you touched a crayfish. Would its body be hard or soft?

Its body would be hard. All arthropods have an **exoskeleton**—a hard outside covering. *Exoskeleton* means "outside skeleton." It holds up the arthropod's body, just as the skeleton inside you holds up your body.

Arthropods also have segmented bodies. For example, spiders have bodies with two main segments. Insects have bodies with three segments. Look at the picture of the insect (the bee). The front segment is the *head*. The middle segment is the *thorax*. What is the last segment?

Right! The last segment is the *abdomen*.

Many arthropods live where people do. Ants, flies, and spiders are some of those arthropods.

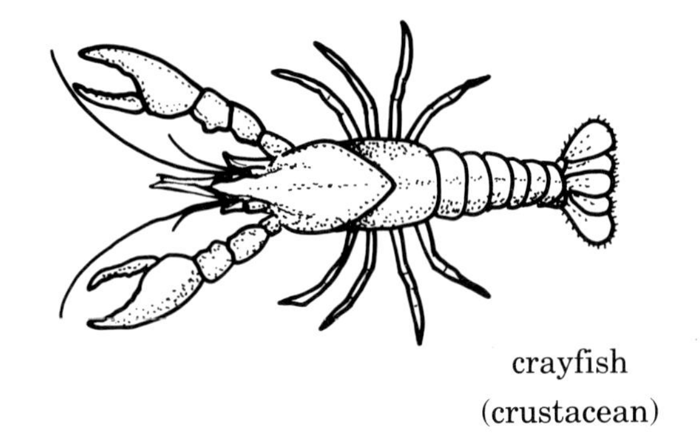

crayfish
(crustacean)

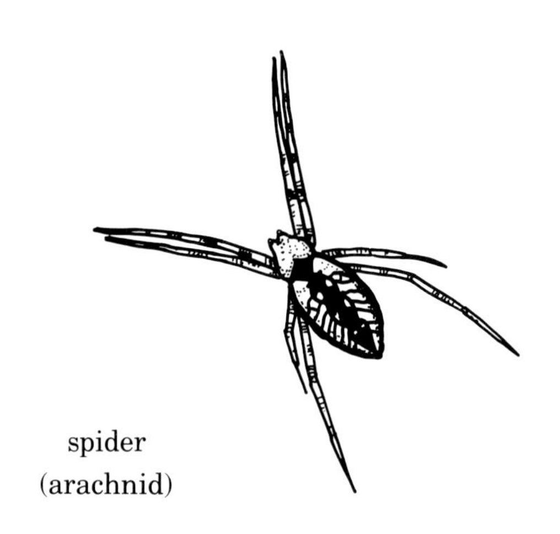

spider
(arachnid)

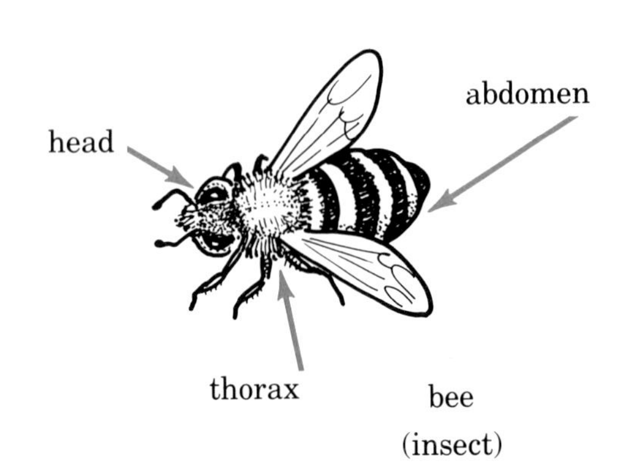

bee
(insect)

Echinoderms

The animals in the *echinoderm* phylum live in the sea. Sea urchins, starfish, and sand dollars belong in that phylum. Those animals all have a covering that is hard and tough. That tough covering is a trait of echinoderms. In fact, *echinoderm* means "spiny skin."

Look at the pictures of the starfish and sand dollar. How are their bodies alike?

Did you notice that both bodies have a five-part pattern? A five-part body pattern is also a trait of many echinoderms.

You can find starfish on rocks or walls in the sea. If you try to pick one up, it will be hard to pull off. Starfish have many small *tube feet* on their arms. Those tube feet are like small suction cups. They stick tightly to things. Tube feet are another trait of echinoderms.

The starfish feeds on mollusks such as oysters and clams. How do you think the starfish gets inside their hard shells?

The starfish wraps itself around a mollusk. It sticks its tube feet on the shells. Then it slowly pulls on the shells until they open a little.

The starfish next pushes its stomach outside its body. It puts its stomach inside the mollusk's shells. Its stomach then digests the mollusk's body.

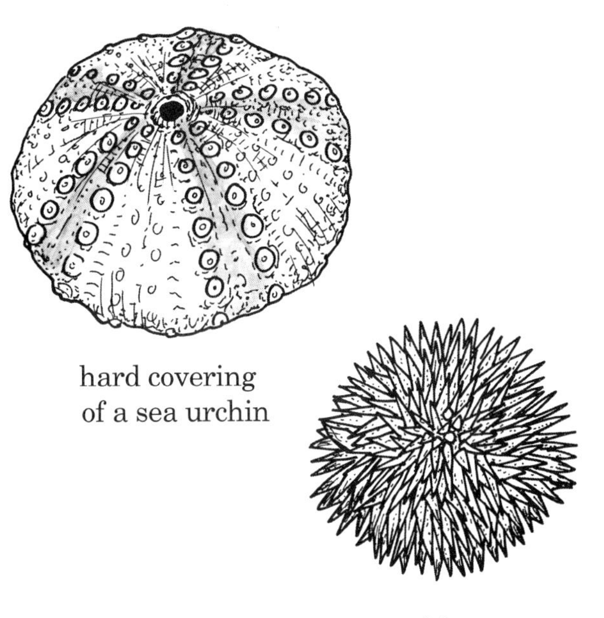

hard covering of a sea urchin

sea urchin

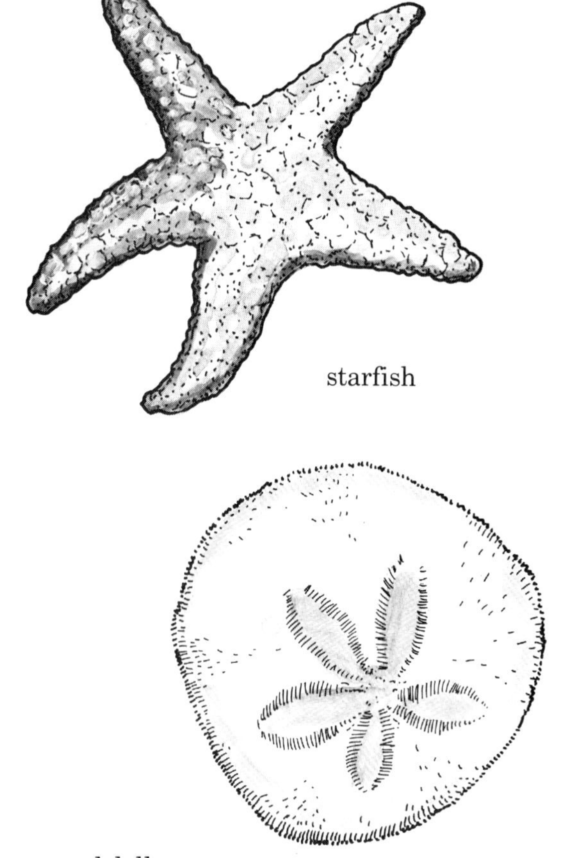

starfish

sand dollar

Vertebrates

Feel the bony ridge that runs down your back. This is your backbone. Animals with backbones belong in the **vertebrate** group. Dogs, birds, and snakes are vertebrates, along with humans.

All vertebrates have a skeleton inside their bodies. This skeleton includes a skull that encloses a brain. The backbone is the main part of a vertebrate's skeleton.

The vertebrate group is not a phylum. It is a smaller group called a **subphylum**. The phylum that vertebrates belong in is the **chordate** phylum. Chordates are animals that have a cord down their backs at some time in their lives. Chordates that aren't also vertebrates are not very common.

Scientists divide the vertebrates into seven classes. The pictures show some of these classes.

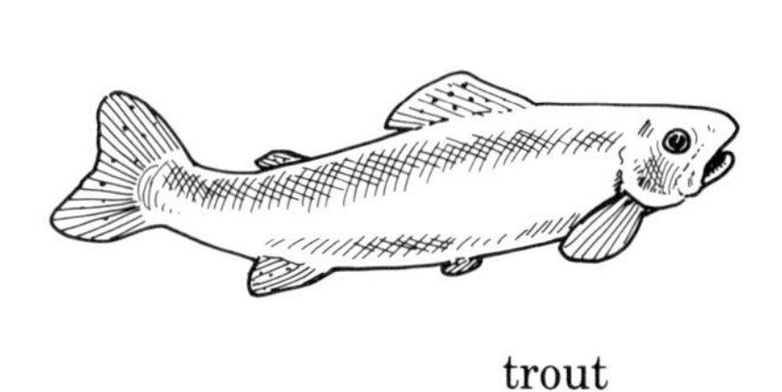

trout

Fishes

There are three classes of fishes. The largest class includes fishes with bony skeletons, like the trout. Another class contains fishes like the sharks, with skeletons made of a flexible material called cartilage. The third class is made up of fishes that have no jaws, like the lamprey.

All fishes live in water and take in oxygen through gills. They use fins to move through the water.

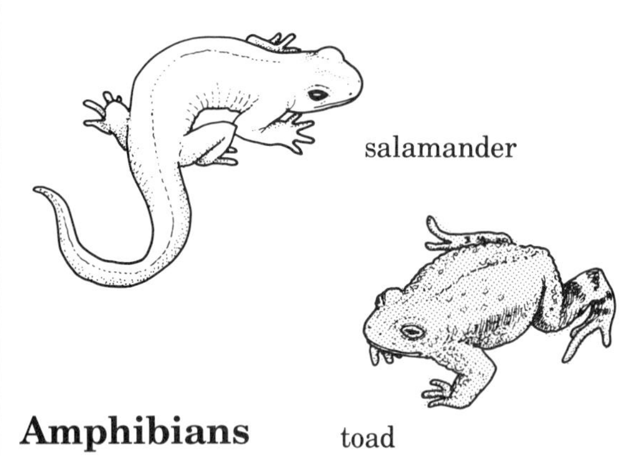

salamander

toad

Amphibians

Many amphibians begin their lives in water. They take in oxygen through gills just like fish. When they are grown up, they live on land and breathe through lungs. They return to water to lay their eggs. The eggs do not have shells.

Most amphibians have skins that dry out easily, so they must stay in moist places.

snake

crocodile

Reptiles

Unlike amphibians, reptiles do not depend on water. Waterproof skin and scales keep them from drying out. Their eggs have a leathery shell that protects them, too. Some reptiles live in water. But all breathe air throughout their lives.

Birds

bluejay

All birds have feathers. No other animals do. Birds also have beaks, wings, and scales on their feet. Most birds use their wings for flight, while others use them for swimming. A few birds cannot fly at all. Birds also lay eggs with hard shells. Their hard eggshells and scales are traits they share with reptiles.

dolphin

cat and kittens

chimpanzee

Mammals

The mammal class is the one you belong in. All mammals have these traits: They are covered with hair. They feed their young milk that is made inside the mother's body. All but one group of mammals give birth to young.

Most mammals live on land, but a few live in the ocean. Dolphins and seals are examples of ocean mammals.

There are many different kinds of mammals. Some are tiny, like the mouse. Some are large, like the elephant. What are some other mammals?

Which Vertebrates?

Which classes do these vertebrates belong in? Take a guess. For each animal, pick a class from the list. Then check your answers. (The right answers are upside down.)

amphibian
bird
fish
mammal
reptile

Answers

1. mammal
2. fish
3. reptile
4. amphibian
5. bird

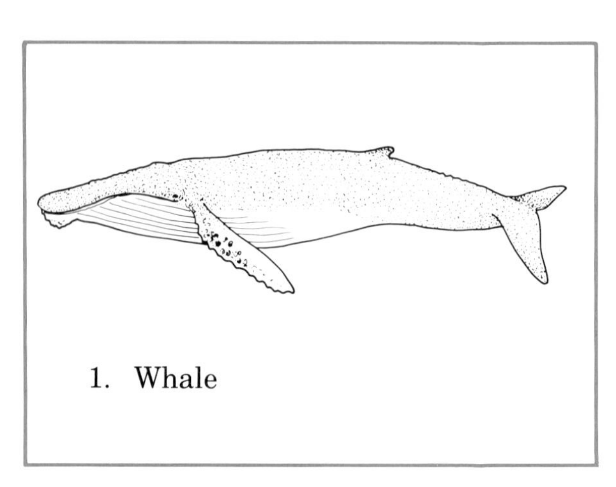

1. Whale

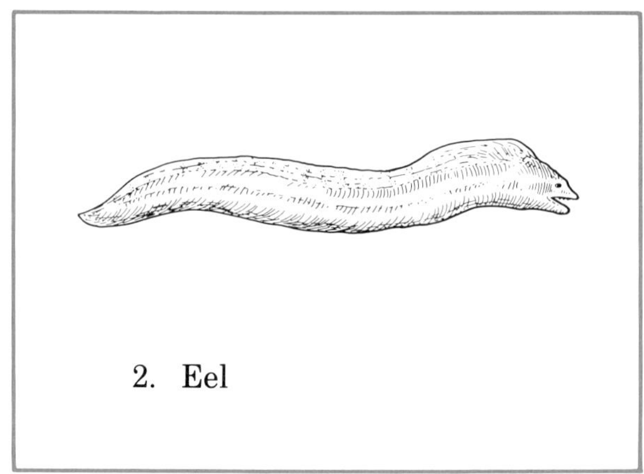

2. Eel

3. Lizard

4. Frog

5. Penguin

Animal Watch

Choose an animal that you can take care of and study. It can be an animal that you keep in school. Or it can be an animal that you care for at home.

Now find out all you can about that animal. Get some books from the library. Talk to experts, such as naturalists and veterinarians. Then answer these questions. On a separate piece of paper, keep a record of what you find out.

1. What animal is it?
2. What phylum or subphylum does it belong in?

sponge	worm	arthropod
coelenterate	mollusk	vertebrate

3. What are some traits of its phylum or subphylum? (Example: Sponges have holes in their bodies.)
4. What class does it belong in? (Examples: insect, spider, mammal, fish, amphibian, reptile, bird)
5. What are some traits of its class? (Example: Insects have segmented bodies.)
6. What other animals are most like that animal? (Example: Dogs, wolves, and coyotes are most like each other.)
7. Draw the animal. Show the traits of that animal's phylum. (Example: Arthropods have jointed legs.) Show a trait of the animal's class. (Example: Mammals are covered with hair.)

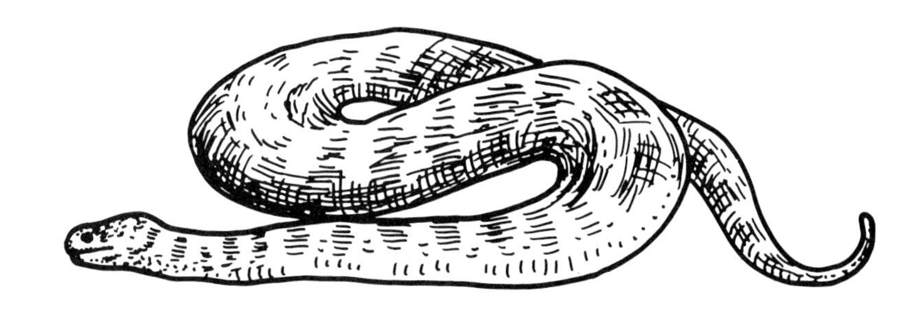

Review

Show what you learned in this unit. What phylum does each of these animals belong in? (Hint: Think about the phylum's traits. One trait is given.) Match each animal on the left with its phylum on the right.

1. spider, bee, crayfish	a. coelenterates (trait: stinging cell)
2. shark, elephant, human	b. segmented worms (trait: digestive system)
3. clam, squid, slug	c. mollusks (trait: soft body)
4. jellyfish, sea anemone, coral	d. arthropods (trait: jointed legs)
5. earthworms	e. echinoderms (trait: spiny skin)
6. sea urchins, starfish, sand dollar	f. chordates and vertebrates (trait: backbone)

Check These Out

1. Invite a veterinarian to speak to your class. Ask him or her what kinds of jobs you can get in animal care.
2. What are organ systems? Find out what organ systems you have. Give a report about one organ system to your class.
3. As a class, get a preserved frog from a science store. Dissect (cut open) the frog and study its different organ systems. Draw one of those systems.
4. Make a poster. Pick one of the seven classes that are part of the vertebrate subphylum. Find and cut out pictures of animals that belong to that class. (Look in magazines and newspapers.)
5. Do you live near an ocean? Find out when the low tides are, and visit a tidepool. Draw or write the names of the animals that you see.
6. Here are more things you may want to find out:
   - Where do zoos get wild animals such as tigers?
   - What are some animals that live in the desert? In the mountains? In the tropics? At the North or South Pole?
   - How are the lamprey and shark different from other fish?

Unit 3

Different Kinds of Bodies

Many different kinds of animals live on Earth. Some are so tiny, you can't see them without a microscope. Some are so large, they would fill up a room. Some live on land. Some live in water. They crawl, walk, swim, or fly.

All those animals have different kinds of bodies. But their bodies must do the same things for all of them. Their bodies must get the things all animals need in order to stay alive.

- How do different animals get oxygen?
- What special parts help them find and eat food?
- How does an animal's body keep it safe?

You'll find out in this unit.

Before You Start

You'll be using the science words below. Find out what they mean. Look them up in the Glossary. On a separate piece of paper, write what the words mean.

1. **predator**
2. **prey**
3. **specialized**

Trout

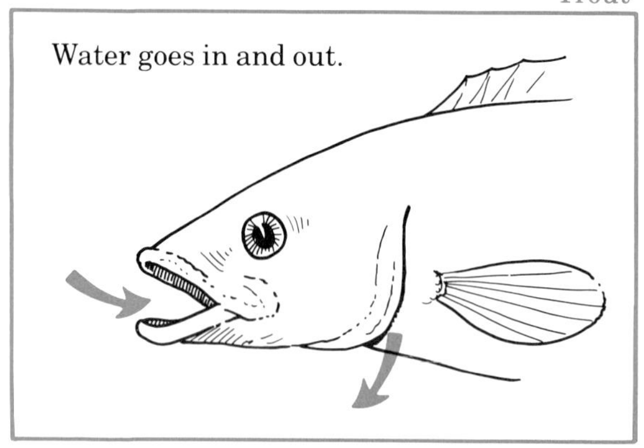

Water goes in and out.

Fishes breathe through their mouths and gill flaps. They have gills that take oxygen out of the water.

Human

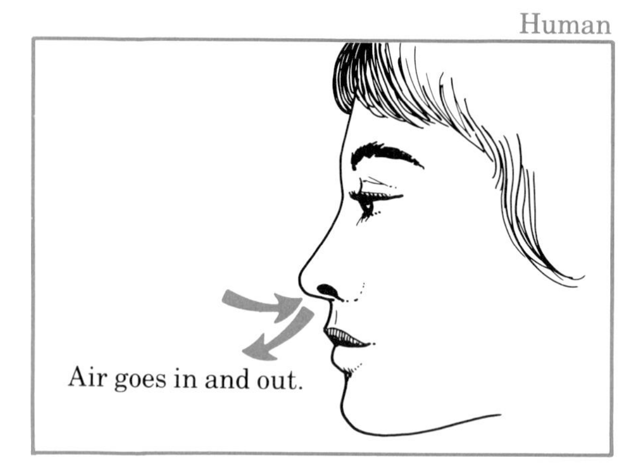

Air goes in and out.

Mammals breathe through their noses. Mammals have lungs that take oxygen out of the air.

Grasshopper

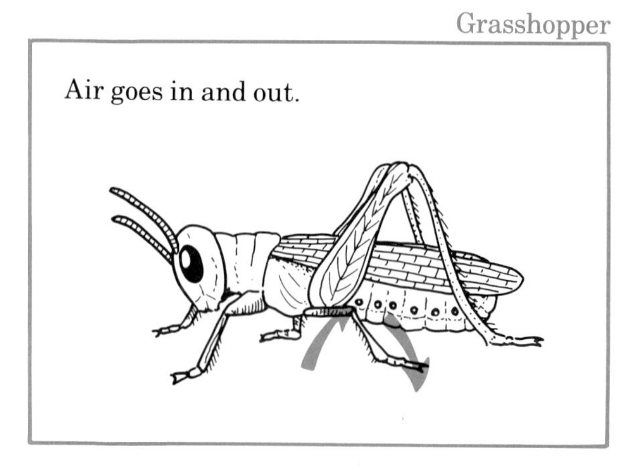

Air goes in and out.

Insects breathe through holes on the sides of their bodies.

Different Ways to Breathe

Animals need oxygen to live. They get oxygen from air or water. So all animals have *specialized* body parts that bring oxygen into their bodies.

Most animals that live in water have parts called *gills*. Water goes into their bodies and passes through those gills. The gills then take oxygen from that water.

Animals that live on land get oxygen from air. Many of them have parts called *lungs*. Those animals breathe air into their bodies. Then the lungs take oxygen from that air.

Some animals that get oxygen from air don't have lungs or gills. Insects are like this. An insect breathes through small holes on the sides of its body. Those holes are connected to long tubes that go inside the insect's body. Air goes through the holes and into the tubes. The tubes take oxygen from that air.

All animals in one class get oxygen the same way. For example, all mammals have lungs and breathe air. All fishes have gills and breathe water. And insects breathe air through holes on the sides of their bodies. The pictures show how animals in those three classes get oxygen.

Now think of a cat, bee, whale, and eel. They are in the same classes as the animals in the pictures on this page. Which animals breathe the same way?

1. Which animal breathes the same way as a cat?
2. Which animal breathes the same way as a bee?
3. Which animal breathes the same way as a whale?
4. Which animal breathes the same way as an eel?

What foods do these animals eat? You can tell by their mouths.

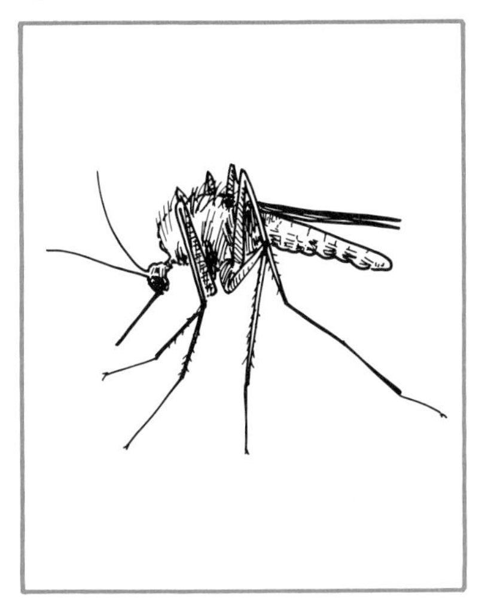

Mosquito

Woodpecker

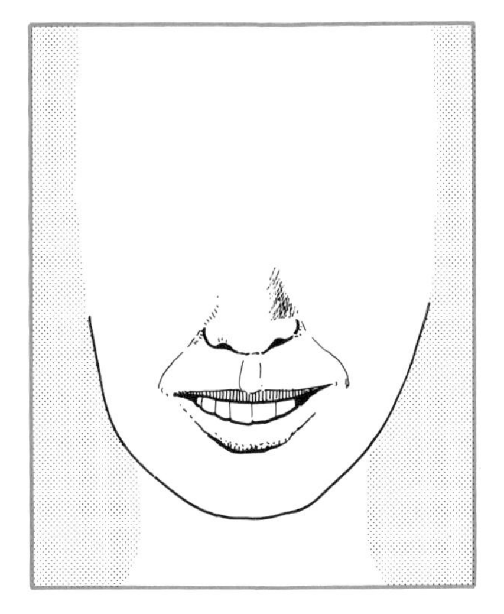

Human

Different Ways to Eat

Animals eat different things. Some eat plants. Some eat other animals. Their mouths have different shapes.

Mosquitoes feed on blood. Their mouths are shaped like a needle. Mosquitoes push their mouths through the skin of an animal such as a human or a horse. Then they suck that animal's blood.

Woodpeckers eat insects that live inside trees. Their *beaks* are hard and pointed. Their tongues are long and sticky. They chip holes in trees with their beaks. Then they push their tongues into the holes to catch the insects.

Some animals have teeth. Meat eaters such as dogs have long, sharp, pointed teeth. Those teeth can hold and tear the flesh of other animals. What other animal has teeth like that?

Animals that eat plants may have sharp, wide teeth. The squirrel is an example. Its teeth can bite and chop up hard stems, roots, nuts, and fruits. What other animal has teeth like that?

Some animals feed on the leaves and seeds of grasses. Those animals have big, flat, wide teeth that can grind those foods. The horse has teeth like that. What other animal does too?

Now think of humans. We eat plants and animals. What kind of teeth do we have?

Built to Get Food

Animals that kill other animals for food are called *predators*. Predators have specialized body parts that help them find, hunt, trap, and catch their *prey*.

Some predators have strong bodies that can move with great speed. Look at the pictures on this page. Which animal can run very fast?

Right! The cheetah can outrun most other animals. It sneaks up on its prey. Then, when it gets close, it dashes out and runs down the other animal.

Some predators have *sense organs* that are very specialized. Those animals can see, hear, or smell their prey when it is far away.

Look at the picture of the bat. The bat hunts for insects. Which of its sense organs do you think are very specialized?

Right! The bat's ears are very specialized.

Bats don't see well. But they have a keen sense of hearing. They send out sounds that are so high, humans can't hear them. Those sounds travel through the air. When the sounds hit the bat's prey, they bounce off and come back to the bat's ears. The bat can tell where its prey is.

Hawks fly high in the sky, so high you almost can't see them. They hunt small animals that live on the ground. They dive quickly out of the sky and grab the animal before it can run for cover. What sense organ helps the hawk high in the sky find its food?

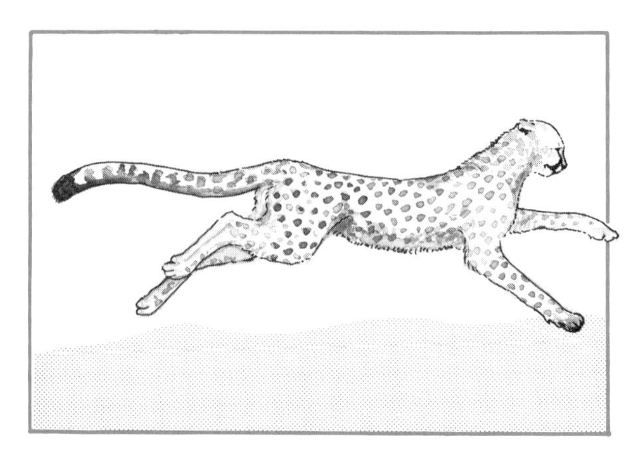

The cheetah is the fastest land animal. It can run up to 70 m.p.h. for short distances.

East Bay Regional Park District

The bat uses sound to find its prey. How are its ears specialized?

East Bay Regional Park District

Where is the lizard in this photograph?

Van Waters & Rogers

Where is the insect in this photograph?

Built to Get Away

Animals have specialized body parts that help them get food. They also have specialized body parts that keep them safe.

Some animals have special tough coverings that protect their bodies. The porcupine is an example. Its skin is covered with sharp quills that can cut the skin on a predator's nose or mouth. When the porcupine is in danger, its quills stand straight out. The porcupine now looks fierce, big, and dangerous. That frightens off the predator.

The turtle also has a tough covering—its shell. How does that shell keep the turtle safe?

Sometimes animals fool other animals by looking like something else. Scientists call that *mimicry.* For example, some moths have big circles on their wings. The circles look like the eyes of a large animal. That keeps the moths' enemies away. Some butterflies look like other butterflies that taste bad. Predators think they are the bad-tasting butterflies and leave them alone.

Some animals use mimicry in a different way. Their color and markings make them look like what they are on. When they stay very still, they look like sand, rocks, twigs, or leaves.

For two examples of mimicry, look at the photographs on this page. One photograph shows a horned lizard. The other shows an insect. Can you find the animal in each picture?

Animal Watch

Watch an animal for a while. It can be an animal you are caring for or an animal that belongs to someone you know. It can also be an animal in a zoo.

Find out what that animal eats. Look at its mouth and other body parts. Try to see what is special about that animal's body. Then, on a separate piece of paper, answer the questions below.

1. How does the animal get oxygen?
2. What kind of food does it eat?
3. What are its mouth parts like? How do those parts help it eat food?
4. What specialized body parts help it get food?
5. What specialized body parts keep it safe from enemies?

Review

Show what you learned in this unit. Finish the sentences. Match the first part of the sentence on the left with the correct words on the right.

1. Every animal has specialized body
2. The animal's specialized body also
3. Gills help certain animals get
4. Lungs help certain animals
5. Animal mouths are specialized to
6. Predators have specialized

a. eat only certain kinds of food.
b. body parts such as sharp teeth.
c. parts that help it get food.
d. oxygen from water.
e. keeps it safe from enemies.
f. get oxygen from air.

Check These Out

1. Find out what *sonar* is, and how bats use sonar. Then write a report about that information.
2. Pick an animal. Study its body. What specialized parts get oxygen? Find and eat food? Keep the animal safe? Find or draw a picture of that animal. Label its specialized parts and write what they do.
3. Collect pictures of animals that show them getting away from their predators. Post some pictures on a bulletin board in class.
4. Some animals, such as the opossum, play dead. Find out which other animals play dead. Tell your class about some of those animals.
5. Visit a zoo or pet store when the animals are being fed. Watch how different animals eat. Then report what you see to your class.
6. Here are more things you may want to find out:
   - What are carnivorous animals? Herbivorous animals? Omnivorous animals?
   - Who was Charles Darwin? What did he do? What is his *Theory of Evolution?*
   - What is *adaptation?* How do some animals adapt to different things?
   - How do corals eat?
   - What is venom? What are some animals that have venom?
   - How do different animals camouflage themselves?

Unit 4

Animal Life Cycles

Animals must be able to produce new animals like themselves. If they couldn't do that, they would die out. There would be no animals like them left on Earth. So all animals have ways to start new life.

Animals have different kinds of babies. Some babies have bodies like their parents. Some do not.

- What different ways do animals produce new animals?
- What kinds of babies do different animals have?
- What different ways do animals grow up?

You'll find out in this unit.

Before You Start

You'll be using the science words below. Find out what they mean. Look them up in the Glossary. On a separate piece of paper, write what the words mean.

1. **develop**
2. **reproduce**
3. **stage**

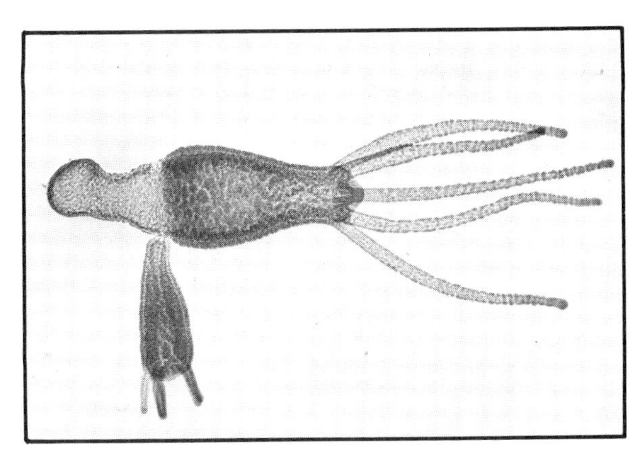

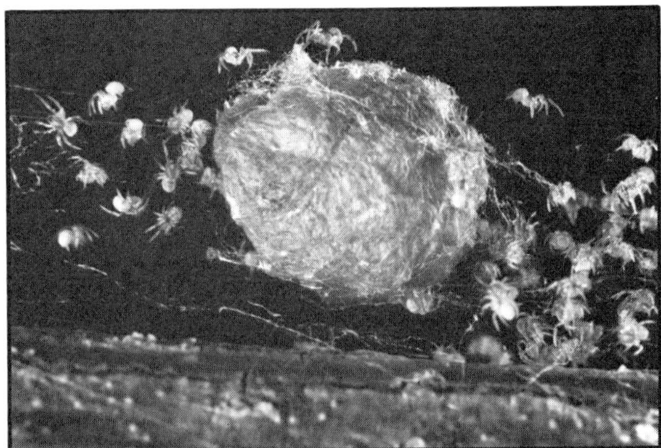

Starting New Life

Animals *reproduce* in different ways. Some of them start new life by *asexual* reproduction. In asexual reproduction, there is only one parent. The new animal is just like the parent.

One kind of asexual reproduction is called budding. A special part—a bud—grows on the parent's body. The bud breaks off and becomes a new animal. Some sponges and coelenterates reproduce by budding.

Most animals reproduce *sexually*. In sexual reproduction, two parents—a male and female—each produce sex cells. The female makes *egg cells*, and the male makes *sperm cells*. When a sperm cell **fertilizes** (joins) an egg cell, new life starts. The egg cell develops into a new animal.

Most animals that live in water fertilize the egg cells outside the female's body. The female pushes the egg cells out of her body. Then the male puts sperm on them.

Most animals that live on land fertilize the egg cells inside the female's body. Most of these land animals then *lay* the fertilized eggs. The young develop inside the eggs. They hatch out of the eggs when their bodies are completely formed.

Some vertebrates *give birth* to living young. These animals keep the fertilized eggs inside their bodies. The eggs slowly develop into young animals. Then the young animals are born. Vertebrates that give birth to live young include most mammals and some reptiles, amphibians, and fish.

Look at the pictures on this page. How does each kind of animal reproduce? Which one gives birth? Which ones lay eggs?

Growing Up

An animal's body goes through several stages as it grows. We call those stages its **life cycle**. Some animals' bodies change completely during the life cycle. Other animals' bodies change just a little.

Humans are animals whose bodies change a little. Our bodies don't change into a different form. They just get bigger as we grow up.

Look at the pictures. They show four main stages of the human life cycle.

Infant: Infants are baby humans. They can't walk or talk. They can't take care of themselves. They don't have teeth and must drink or eat soft foods.

Child: An infant grows bigger and becomes a child. Children can walk, run, and talk. They have teeth and can feed themselves. But they still need to be taken care of.

Adolescent: The child grows into an adolescent—a teenager. The person's body begins to develop sexually.

Adult: When adolescents finish growing, they are adults. They now can take care of themselves and their young.

Many other animals have life cycles like humans. They change just a little. What are some of those animals?

1. eggs

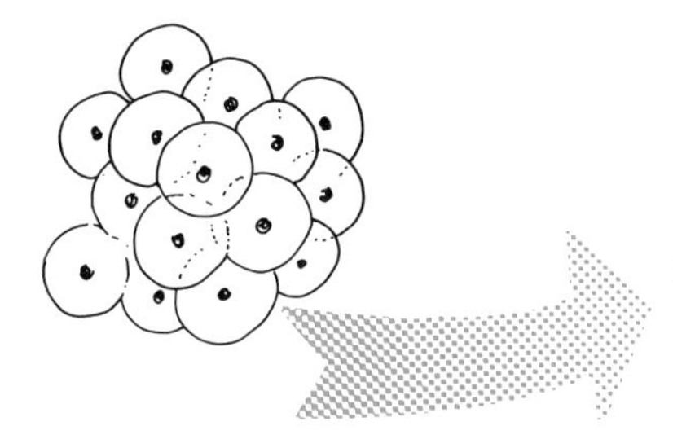

2. just-hatched tadpole

3. older tadpole with legs

4. adult frog

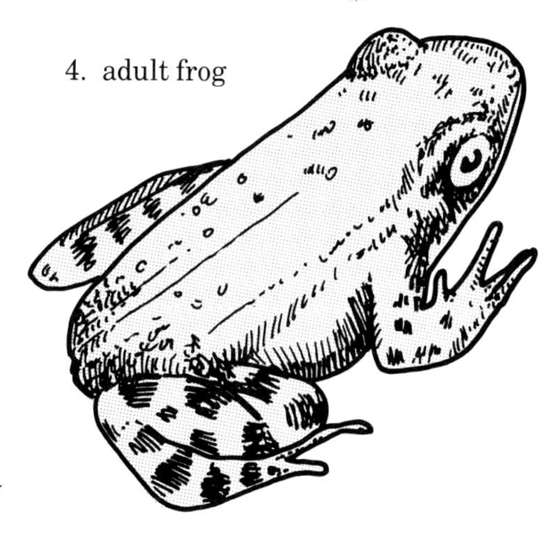

Changing Bodies

Some animals' bodies change completely during the life cycle. Those animals have one kind of body when they are born and another kind of body when they are fully grown. We call that change *metamorphosis*.

Frogs go through metamorphosis. The pictures on this page show how the frog's body changes. Look at the pictures, and then answer these questions.

1. What is the first stage of a frog's life cycle?
2. Frogs hatch from eggs. When the eggs hatch, the baby animal looks and acts like a tiny fish. It swims in water and breathes with gills. It eats tiny plants. What is the animal called at this stage?
3. As the tadpole grows bigger, its body begins to change. New parts start to grow on its body. What are those parts?
4. The tadpole grows into an adult frog. Its body changes completely. The animal now lives on land. It eats insects. Its gills have changed to lungs. How else has the frog's body changed?

Another Kind of Metamorphosis

You've learned that frogs have bodies that change completely. Butterflies and most insects also go through metamorphosis. But their metamorphosis is different in this way: It has one extra stage called the *pupa stage*.

The pictures on this page show the life cycle of a butterfly. Here are the stages:

Egg: The female butterfly lays eggs on the leaves of certain plants.

Larva: When an egg hatches, it becomes a larva. (The butterfly's larva is also called a caterpillar.) The larva stage is like the frog's tadpole stage. Look at the picture. What does the larva look like?

Pupa: The larva spends all its time eating leaves. It grows bigger. Then one day it starts to change. Its body forms a hard shell—a *cocoon*. The animal is now in the pupa stage. Inside the cocoon, the larva's body slowly changes into a butterfly's body.

Adult: Now the butterfly's body is completely formed. The butterfly breaks out of the cocoon. How is its body different from the larva's?

The life cycle of a butterfly

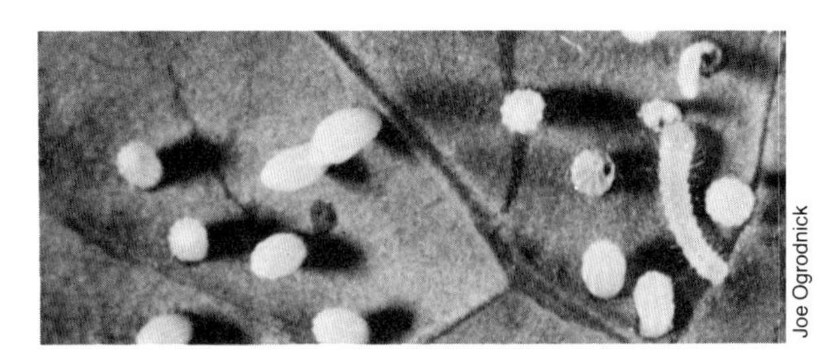
Joe Ogrodnick

1. Egg

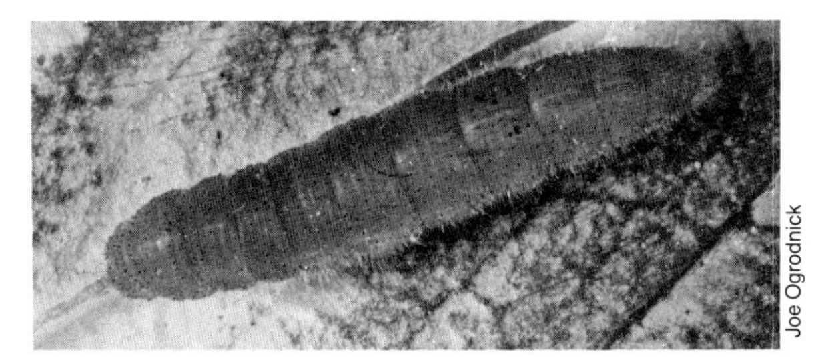
Joe Ogrodnick

2. Larva

Joe Ogrodnick

3. Pupa

Joe Ogrodnick

4. Adult

Review

Use some of the words you have learned in this unit to answer the questions. The words you'll need are listed below.

adult	life cycle	reproduce
budding	mammal	reptile
develop	metamorphosis	sex cells

1. What's another word for "a full grown animal"?
2. What do you call it when an animal's body changes completely?
3. What is one way that sponges reproduce?
4. What are the special cells males and females make?
5. What does an animal's body go through as it grows from a baby to an adult?
6. What is one type of animal that hatches out of an egg that its mother lays?
7. What animal gives birth to living young?
8. What is a word meaning "to produce another animal"?
9. What is a word meaning "to grow"?

Check These Out

1. Put butterfly or moth eggs into a jar. When they hatch, feed them leaves. Cover the jar. (Be sure to make small holes in the cover for air.) Watch each larva change to an adult.
2. Get fish that lay eggs, such as angel fish. Get fish that give birth, such as guppies. Put them in the same tank. Watch their reproductive cycles.
3. Get pictures that show members of your family, from babies to adults. Put the pictures on a bulletin board. Group them so they show the four main stages of the human life cycle.
4. Kangaroos care for their young in a very interesting way. Find out about kangaroos.
5. Here are more things you may want to find out:
   - How do different animals feed their young? Protect their young?
   - What animal has the shortest life cycle? The longest life cycle?
   - How long does a mother carry these animals before they are born: cat, elephant, human, whale?
   - How long do these eggs take to hatch: spider, chicken, lizard, turtle?

Unit 5

Animal Lifestyles

You learned that animals have different kinds of bodies. Those bodies are specialized to help the animals keep alive.

Animals also have different ways to live and act. Those ways are also specialized to help the animals keep alive.

Animals have many, many ways to do things. Scientists are still finding out about them. Special scientists watch the way animals live in the wild. Other scientists study the way animals act.

- What do scientists say about how animals act?
- What are some ways animals live?
- What are some ways animals care for their young?

You'll find out in this unit.

Before You Start

You'll be using the science words below. Find out what they mean. Look them up in the Glossary. On a separate piece of paper, write what the words mean.

1. **behave**
2. **behavior**
3. **survive**

How Animals Behave

The way an animal behaves is the way it acts. An animal behaves pretty much the same way as other animals in its **species**. (A species is a group of just one kind of animal.) For example, all dogs have certain ways of acting. And all the birds in a species build the same kind of nest. Each kind of animal behaves in ways that help it survive.

Scientists who study animal behavior think this: When an animal is born, it already knows how to behave in certain ways. Scientists call that kind of behavior **instinctive behavior**. For example, spiders seem to know how to spin their webs. And human babies don't need to be taught how to suck milk. What do you think is another example of instinctive behavior?

Scientists also think that an animal must learn how to do certain things. They call that kind of behavior **learned behavior**. For example, lion cubs seem to learn how to hunt by watching their mothers. And humans learn to talk by watching and listening to other people. What is another example of learned behavior?

Some animals can be trained to behave in special ways. Dogs can be trained to guide blind people. Elephants can be trained to move heavy objects such as logs. Horses can be trained to carry people. What is another example of trained behavior?

Now think of an animal you are caring for. Could it be trained to do something? What could it do?

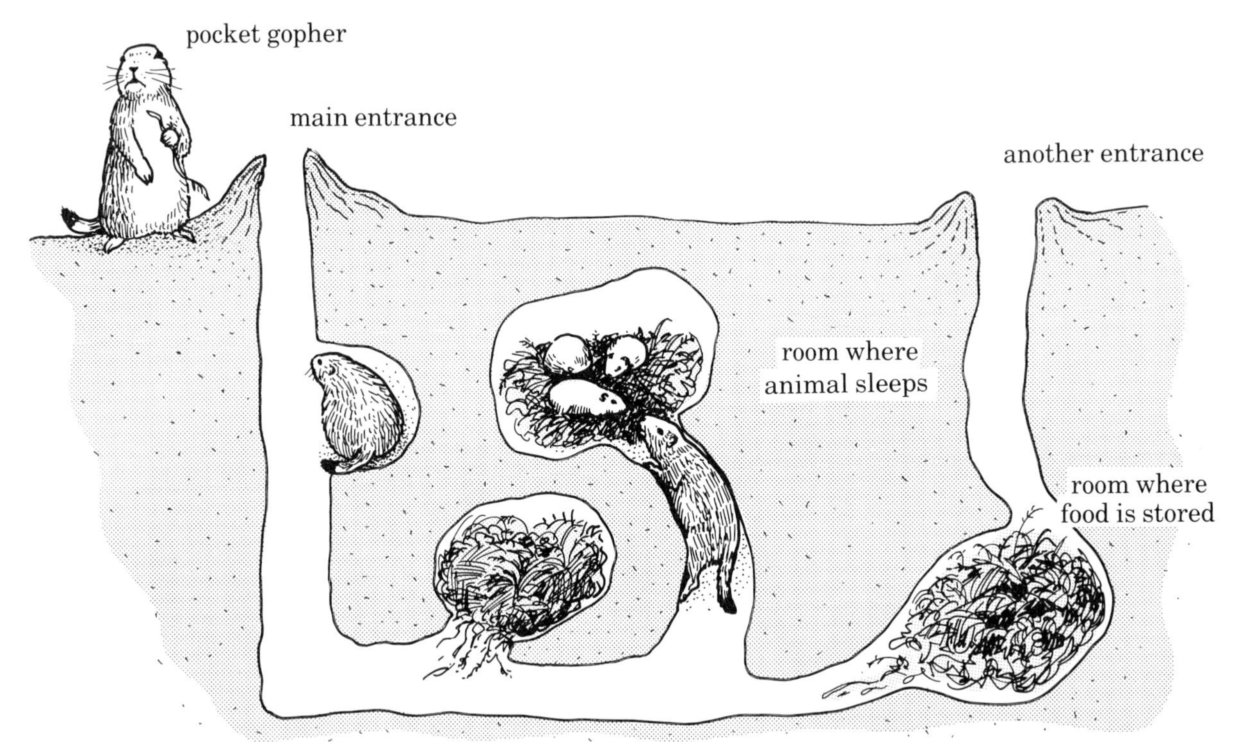

Pocket gophers build homes with many rooms. They use the rooms for different things.

Different Ways to Live

Think of a fly. Now think of a honeybee. Both insects have different ways to live. One moves from place to place. And the other always goes back to a home. Which animal moves from place to place?

Right! The fly moves from place to place.

Many animals are like the fly. They don't have a special place that they always go to. They move around, looking for food. What's another animal that moves from place to place?

Other animals live in homes. Some of them live in places such as caves and holes in trees. Those are animals that find their homes. The bear is one example of an animal that does that. What's another animal that finds its home?

But many animals make their homes. For example, beavers build *lodges* with twigs and branches. Certain termites build huge dirt *nests*. Certain fish build *dens* with tiny stones. And animals like the pocket gopher dig *burrows* in the ground. What's another animal that makes its home?

Animal homes sometimes have many rooms. Humans make those kinds of homes. So do pocket gophers. Look at the picture on this page. How is the pocket gopher's home like a human home?

The rat eats insects. The snake eats rats. What do you think eats snakes?

What's for Dinner?

All animals must eat. So many animals spend a lot of their time looking for food.

Some animals move over large areas searching for food. Deer and bears are some of those animals. The deer searches for the kinds of plants it likes to eat. The bear looks for berries and other animals.

Animals that eat other animals must hunt or trap their prey. What are some animals that hunt for food?

Some animals hunt in groups. They work together to catch their prey. Several of them will run down their prey and kill it. Then they share it. Wolves and lions are examples of animals that hunt in groups.

Animals such as spiders wait for their prey to come to them. Then they trap it. Spiders' traps are sticky webs that they spin. Insects fly into the webs and are caught. The spiders then eat those insects.

Here's another way some animals get food: They *raise* it. For example, one species of ant "plants" a kind of mushroom in its nest. That mushroom is food for the ants. Another species of ant keeps aphids. Aphids are small insects. The ants protect and feed the aphids. The aphids make a sweet liquid that the ants eat.

What's another animal that raises plants and animals for food?

Bringing Up Baby

Suppose a baby snake just hatches from its egg. And suppose a baby bird does too. Can that snake take care of itself? Can that bird take care of itself?

The snake can survive as soon as it hatches. But a baby bird can't.

Some animals can survive by themselves when they are babies. Reptiles, insects, and most animals that live in water are like that.

But other animals must be cared for by adults until they are big enough to survive. What are some of those animals?

Adults care for their young by feeding them. Some animals hunt for food for their babies. They bring the food back to their young. Mammals make milk in their bodies and feed it to their young.

Adults also care for their young by protecting them. They hide their young in places that are hard to find. They attack other animals that come near. Some animals also do this: They trick other animals into following them away from their young.

Adults care for their babies until the young animals are almost grown up. A deer mouse takes about six weeks to grow up. So it stays with its mother for six weeks. A raccoon stays for up to a year. How long do humans care for their young?

Frank M. Blake, U.S. Fish & Wildlife Service

A mother opossum carries her babies on her back. She keeps them safe. The young opossums leave their mother when they are about four months old.

Carolina Biological Supply

The large insect in the center is the queen bee. She lays eggs. The bees around her are worker bees. They feed the queen and take care of her.

Getting Along

Many animals don't have much to do with other animals. They live alone most of their lives. They are with other animals in their species only when they mate and raise young.

But some animals live close to others like them. They form large groups. For example, fish swim in *schools*. Birds form *flocks*. And wolves form *packs*.

Some animals also form a special kind of group called a *society*. The animals in a society act with each other in certain ways. Sometimes they help each other. Sometimes they do certain jobs.

Honeybees are insects that live in a society. They live in a *hive*. Each bee does a certain job. Some bees gather and make food. Others clean and protect the hive, and so on.

Humans live in a society. So do most monkeys. Watch the monkeys in a zoo. See how they behave with each other.

Sometimes an animal lives with an animal of another species in a special way. Scientists call that **symbiosis**. Here's an example of symbiosis: A kind of crab always carries a sea anemone on its shell. When the crab eats, some of its food also feeds the anemone. When an enemy comes near, the anemone protects the crab by using its stinging cells.

There is one animal that lives with many different kinds of animals. That animal cares for them and keeps them as pets. What animal is that?

Animal Watch

Study a wild animal. It could be one that lives outdoors, such as raccoons, fish, or birds. It could be one that lives in your school building, such as cockroaches, spiders, or mice. It should not be a pet or an animal that lives in a zoo.

Answer these questions about that animal. On a separate piece of paper, keep a record of what you find out.

1. What phylum (or subphylum) does that animal belong in? (If you're not sure, look through Unit 2 of this section.)
2. Where does that animal live?
3. What does it eat?
4. How does it get its food?
5. How does it behave when it gets its food?
6. How does it behave when it sees or notices you?

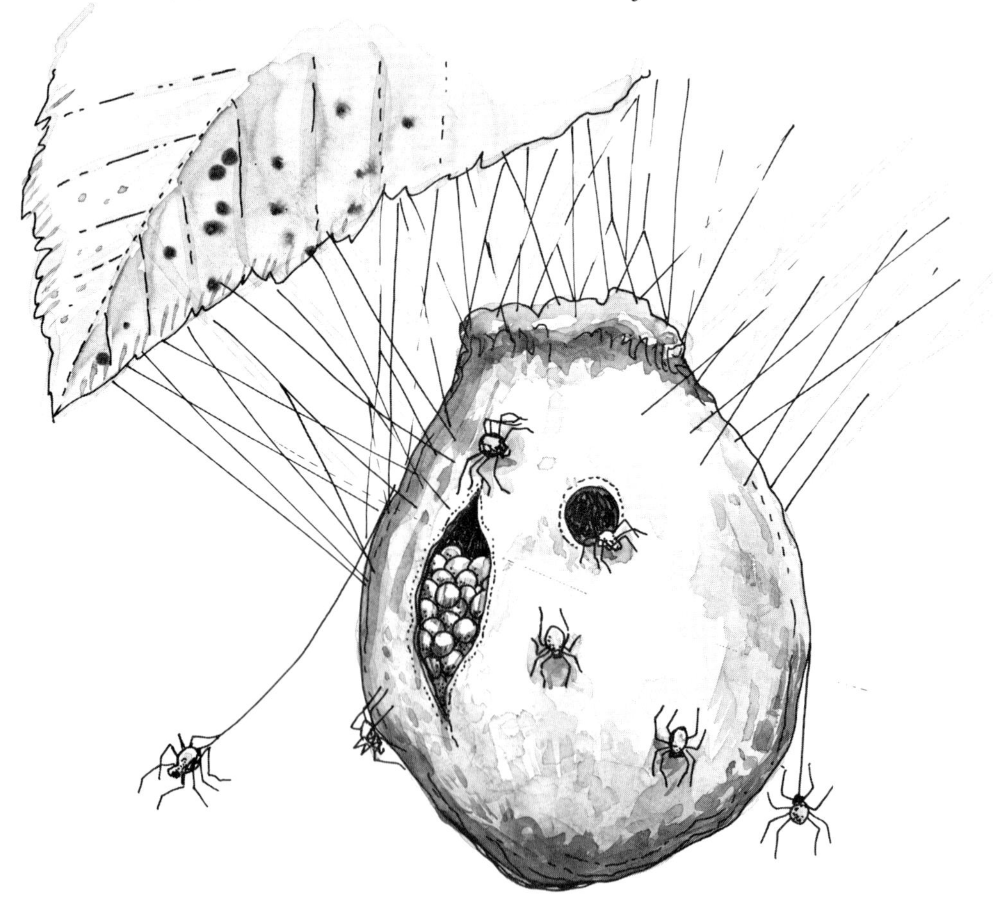

Review

Show what you learned in this unit. Use the words below to answer the questions.

hunting	survive
instinctive behavior	societies
learned behavior	move from place to place

1. Most animals know how to do certain things when they are born. What is that behavior called?
2. Animals have to be taught how to do certain things. What is that behavior called?
3. Some animals live in homes. What do animals like the fly do?
4. When animals form groups to catch their prey, what are they doing?
5. What can some animals do on their own as soon as they are born?
6. What are the special groups that humans, bees, ants, or certain monkeys live in?

Check These Out

1. Many groups of birds *migrate* (move) to warmer climates at a certain time of year. Where do some birds migrate to? When do they migrate? Give a report to your class.
2. Certain animals *hibernate* during cold seasons. Find out what *hibernation* is. What animals hibernate?
3. Aesop's fables are stories about animals. Read one of those stories and tell your class about it.
4. Rudyard Kipling wrote many funny stories about how animals got their body parts. Find out one story. Draw a picture strip that tells the story.
5. Find out about honeybees. How do they form a new beehive? What is life like in the beehive? What do the queen bee, workers, and drones do?
6. Here are more things you may want to find out:
   - What are some ways different animals find their mates?
   - Why do scientists believe dolphins have their own language?
   - Why did astronauts take bees with them on a space shuttle flight? What did the astronauts learn?
   - Who is Jane Van Lawick Goodall? What animals has she studied? What has she found out?

Unit 6

Animals in Danger

Animals have specialized bodies that help them survive. They have specialized behaviors that also help them survive.

But sometimes a whole species of animals has trouble surviving. Large numbers of those animals die. The ones that live can't reproduce fast enough. Soon there are fewer and fewer of those animals. And then one day, there are no more.

- What do animals need in order to survive?
- How does a species survive?
- What causes a species to disappear?

You'll find out in this unit.

Before You Start

You'll be using the science words below. Find out what they mean. Look them up in the Glossary. On a separate piece of paper, write what the words mean.

1. **balance**
2. **endangered**
3. **environment**

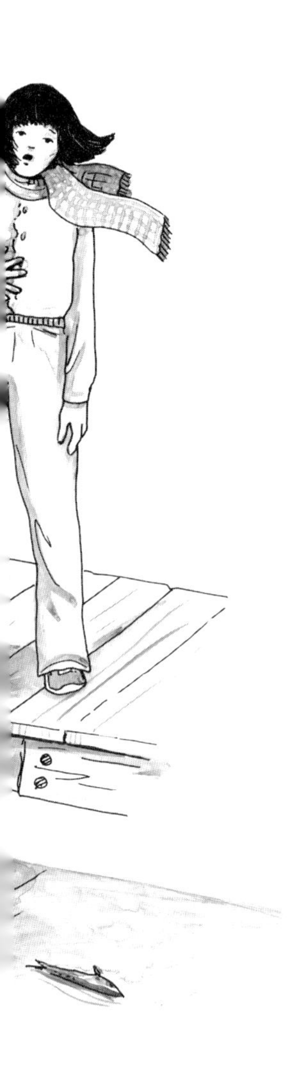

Upsetting the Balance

Imagine a pond in a forest clearing. Many animals live there and *interact* with each other. Fish and frogs eat insects in the water. Snakes and birds eat the fish and frogs.

The animals also interact with the plants that grow in and around the pond. Beavers cut down young trees to build lodge homes to live in. The insects in the water eat water plants.

The many interactions between the living things in and around the pond show that all the living things depend on each other. Together, they form an **ecosystem**.

Ecosystems are normally balanced. This means that the number of each kind of living thing stays about the same over time. For example, the number of frogs eaten by birds and snakes equals the number of tadpoles that grow up to be new frogs.

But the balance of ecosystems can be upset very easily. Humans often upset this balance by accident.

For example, what would happen if the forest around the pond were cleared for houses? The snakes and birds would be scared away or have no place to live. The fish would not get eaten. After a while there would be too many fish. They would eat so many insects that the insects would die off. Without insects to eat, the fish would then die off. And, the water plants would grow out of control without insects to eat them. Because of one change in the ecosystem, every other part of the ecosystem would be affected. In this way, upsetting the balance of an ecosystem can hurt many kinds of living things.

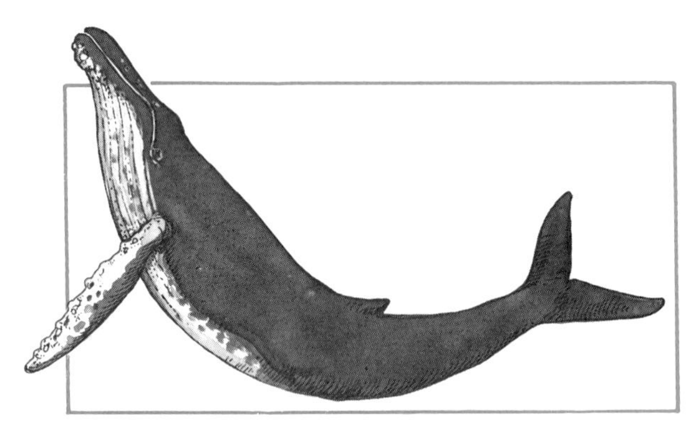

Humpback whale

L. David Mech, U.S. Fish & Wildlife Service

Timber wolf

Mike Lockhart, U.S. Fish & Wildlife Service

Bald eagle

William S. Keller, National Park Service

Grizzly bear

Endangered Species

You will never see a live passenger pigeon. That species is **extinct**. But about a hundred years ago, flocks of those birds lived in the United States. Every year, human hunters killed thousands of them. And the species died out.

A species can die out if it is hunted too much, the way the passenger pigeons were. It can die out if its environment is suddenly changed. And it can die out if its environment becomes polluted with harmful chemicals.

Scientists fear that many species today are endangered. In an endangered species, too many animals die. Not enough are reproduced.

The animals shown on this page are all endangered. The sentences below tell why. Read the sentences. Then name the animals you think they describe.

1. This big mammal lives in the ocean. Humans hunt it in huge ships. Its meat is used to feed humans and pet animals.
2. This bird is the symbol of the United States. Human hunters and ranchers killed many of this bird. Chemicals such as pesticides are polluting its environment.
3. This big mammal is one of Earth's meat eaters. Human hunters kill it for sport. Humans are also changing its environment quickly.
4. This mammal lives in packs. It is related to the dog. Humans kill it for its fur. They also kill it because it is a predator. Humans are taking over its environment.

Animal Watch

Find out about an endangered species. Choose one from the pictures on the other page. Or choose another endangered species that you have heard about.

Learn as much as you can about the animal in that species. Read news stories or books. Talk to people in conservation groups. Talk to scientists, veterinarians, biologists, science teachers, and so on. Find the answers to these questions. On a separate piece of paper, keep a record of what you learn.

1. What is the name of the animal?
2. What phylum does it belong to?
3. What are some body traits of the animal's species?
4. What are some behavior traits of the species?
5. How does the species protect itself?
6. How does it get its food?
7. What kind of environment does it live in?
8. Why is it endangered?

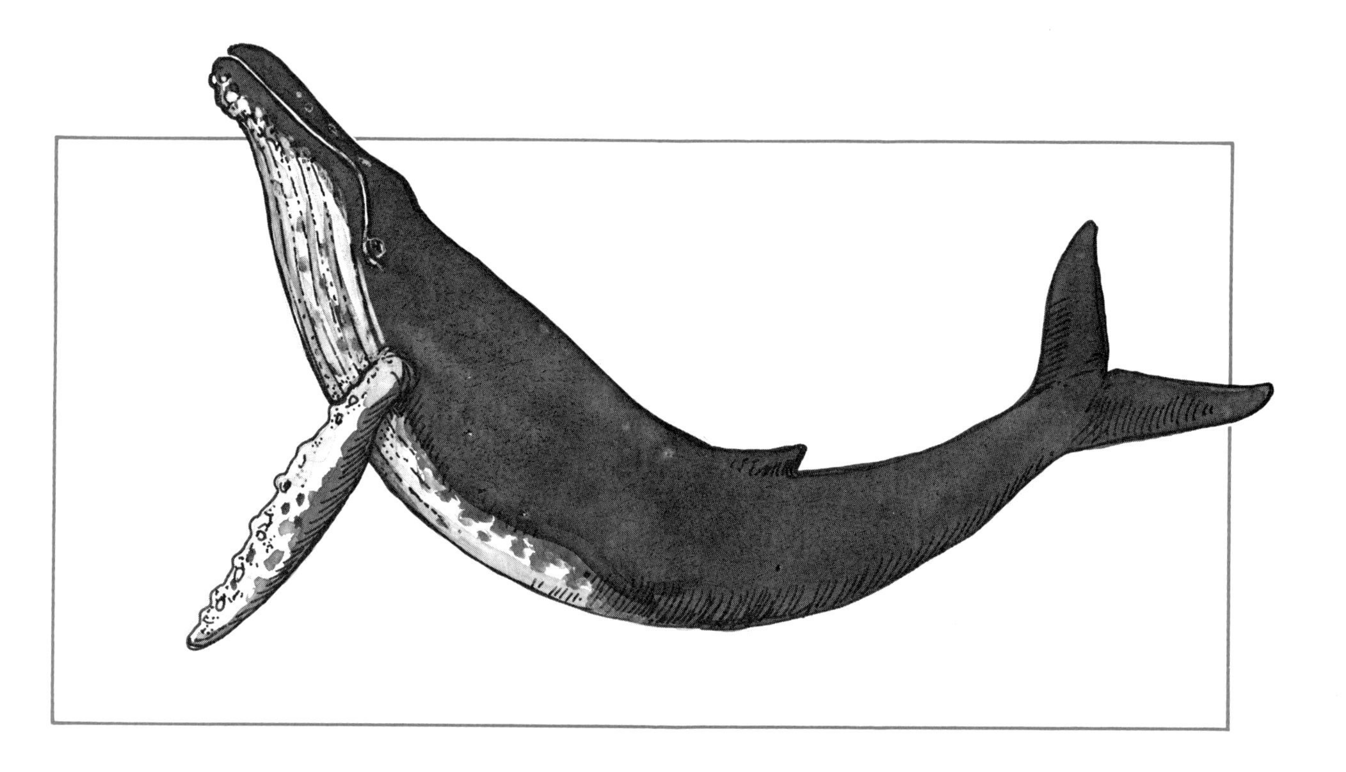

Show What You Learned

What's the Answer?

Choose the correct answer for each question. There may be more than one correct answer to a question.

1. What are some traits of all animals?
   a. They must eat food.
   b. Most can move freely.
   c. All have a backbone.
2. What is a trait of a phylum?
   a. Has an outside skeleton
   b. Classifies animals
   c. Has fur or hair
3. How do specialized body parts help an animal? They
   a. bring oxygen into its body.
   b. keep it safe from predators.
   c. help it find food.
4. How do animals reproduce?
   a. Budding
   b. Laying eggs
   c. Giving birth to young
5. What are animal behaviors?
   a. Building a nest
   b. Having a backbone
   c. Learning to talk
6. How is a species endangered?
   a. It moves freely.
   b. Its environment is polluted.
   c. It is hunted too much by predators.

What's the Word?

Give the correct word or words for each meaning.

1. To put into groups
   C ________
2. The certain ways an animal looks or acts
   T ________
3. To keep alive
   S ________
4. An animal that kills other animals for food
   P ________
5. Body changes from a baby to an adult
   L ________ C ________
6. To act in a certain way
   B ________

Congratulations!

You've learned a lot about animals. You've learned

- About the main kinds of animals
- How animals are alike and different
- How animals survive
- And many other important facts about animals

HUMAN SYSTEMS

What kind of living things are humans? How are we like other living things? What systems keep us alive? How do the human systems work? In this section, you'll learn many facts about the human systems. And you'll learn how those systems keep us alive.

Contents

Introduction

Picture this:

You're fixing a roof. You wipe sweat from your face. You take in a deep breath. You are tired!

You've been working all day. You carry heavy things up and down a ladder. You pull out nails. And you hammer new parts into the roof.

Although you're tired, you keep working. You think: This work is hard, but it's fun. You talk and laugh with other workers. And you feel that your whole body is stronger because of the hard work.

Now you hammer the last nail into the roof. You look at your work and smile. You did a great job!

You can do all those things because your body is made up of different parts. Those parts are called *organ systems*. Those organ systems are always working to keep you healthy and strong. They work so you can climb ladders and use hammers. So you can think and talk, smile and breathe, and do so many other things.

In this section, you'll learn about some organ systems. You'll learn how they work to get what your body needs. You'll learn how all the organ systems work together to keep you alive. And you'll learn just how wonderful your body is!

Unit 1

You're Alive!

Think about this: You do many different things. You work. You play. You talk, think, and laugh. Many parts of your body work together so you can do those different things. Your body parts are always working. They keep you alive!

- What are the parts that make up your body?
- How do these parts work together?

You'll learn the answers in this unit.

Before You Start

You'll be using the science words below. Find out what they mean. Look them up in the Glossary that's at the back of this book. On a separate piece of paper, write what the words mean.

1. **cell**
2. **divide**

What's Your Body Made Of?

The smallest living thing in the world is a *cell*. Cells make up everything that is alive.

You are alive. What are the smallest living things that make up your body?

Right! They are cells!

Cells make up every part of you. They make up your eyelids. They make up your hair and your bones. Name another part of your body that's made up of cells.

You have many different kinds of cells. A group of the same kind of cells is called a **tissue**. The cells that make up a tissue all look alike. They work together to do the same job. For example, your skin is made up of tissues. All the cells in the tissues look alike, and they all do the same job.

Your body has different kinds of tissues. In some parts of your body, different tissues work together. When that happens, the tissues form an **organ**.

You have many organs. Your heart is an organ. What's another organ?

Certain organs work together to do one big job in your body. When organs work together, they make up an *organ system*. Your body has several organ systems. And each one does a special big job. One of your organ systems works so you can breathe. What's another big job that an organ system does?

Look at the pictures. Answer the question under the pictures.

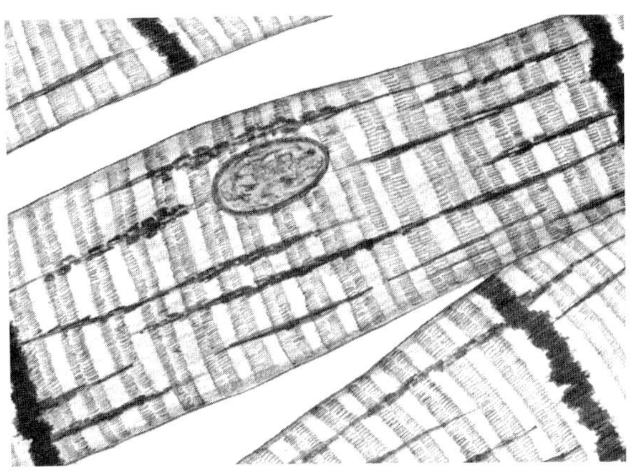

Cells make up . . .

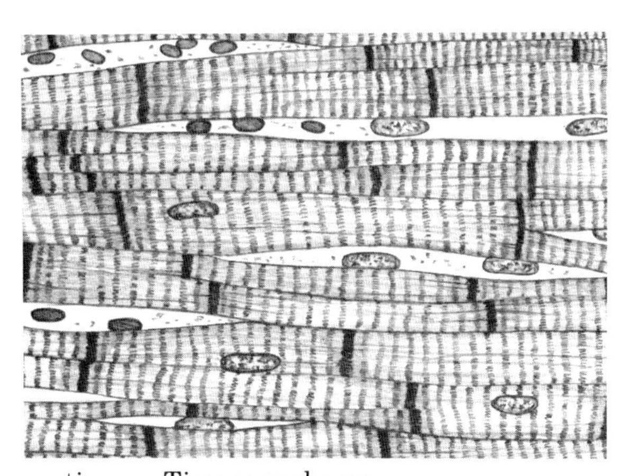

. . . tissues. Tissues make up . . .

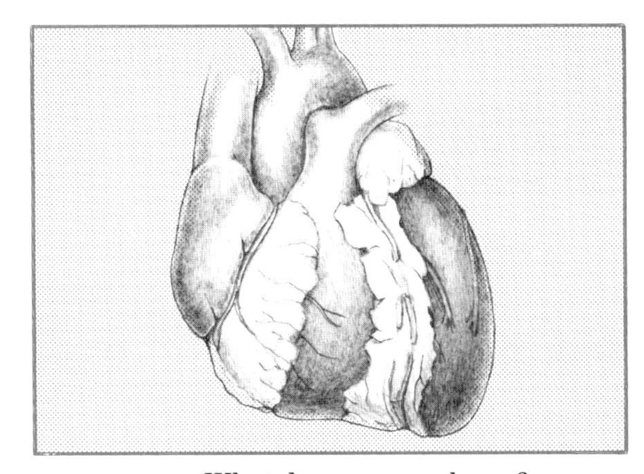

. . . an organ. What do organs make up?

Close-Up of a Cell

See what a cell from your body looks like. You'll need a microscope and these materials:

- One dropper with water in it
- One glass slide
- One toothpick (the kind that has a flat end)
- Some methylene blue stain
- One cover slip

1 Put a drop of water on the center of the glass slide.

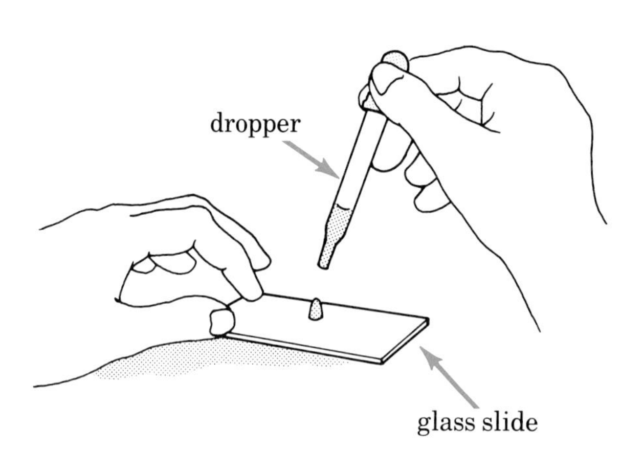

2 Gently scrape the inside of your cheek with the toothpick. (Use the flat end.)

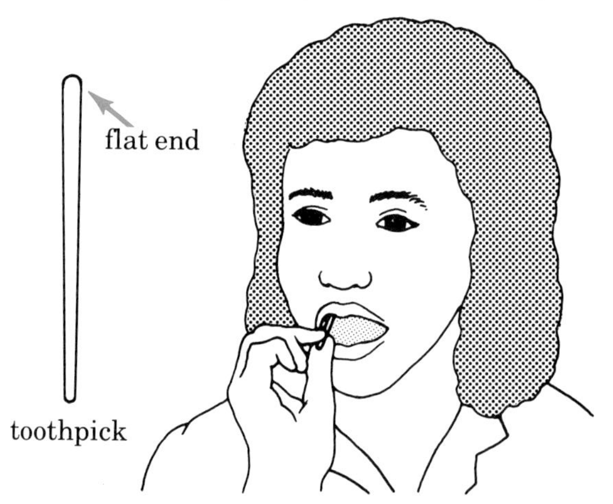

3 Put the flat end of the toothpick in the water. Stir it a few times.

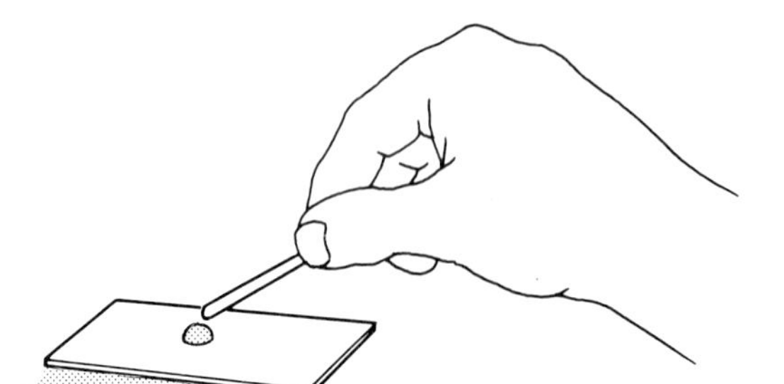

4 Add a drop of stain.

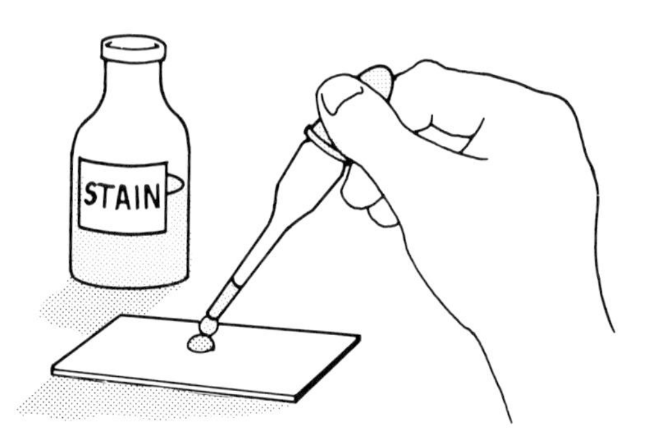

5 Hold the cover slip by its edges. Hold it at one end of the slide, the way this picture shows.

Slowly move the cover slip until it is over the water.

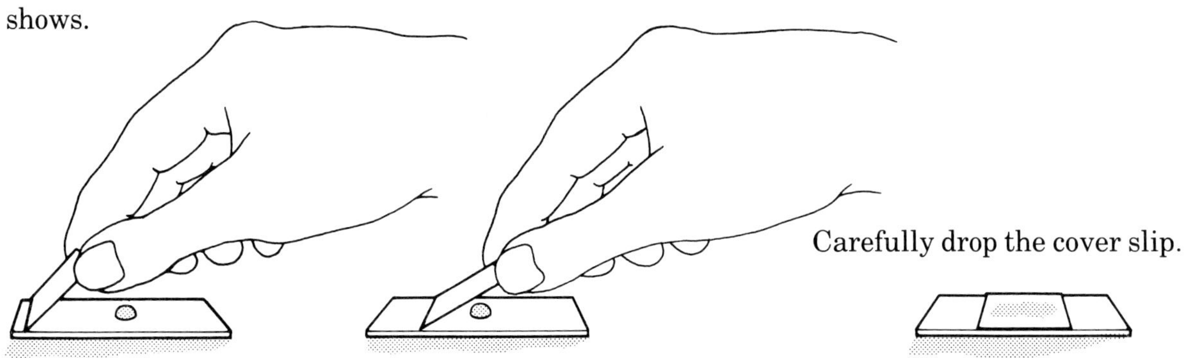

Carefully drop the cover slip.

---

6 Put the slide in the microscope. Turn to low power. Look into the lens. Focus it. You should see several purple dots. They are cheek cells.

You should see something like this under low power.

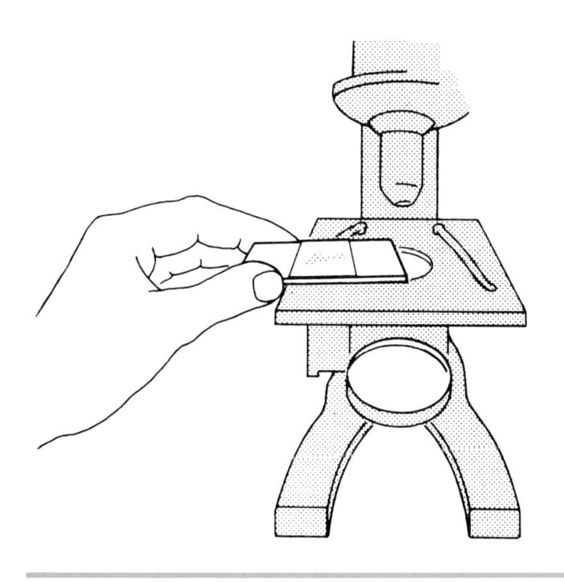

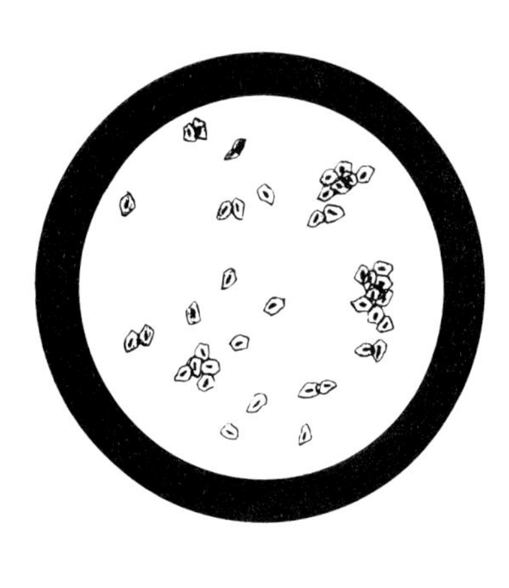

---

7 Keep looking at the cheek cells. Move the slide around until one of the cells is in the middle of the lens. Turn to high power. Focus the lens. You should now clearly see one cell. On a separate piece of paper, draw that cell.

Cells Divide

You've probably seen pictures of yourself as a baby. Remember how small you were? Now your body is much bigger.

Something happened to the cells in your body so that you got bigger. What do you think happened?

The cells in your body *divided*. Each cell split into two cells. That means the cells made more cells. As more cells were made, you grew.

How does a cell divide? It makes a copy of itself, and splits in half. It becomes two cells. The pictures show a cell dividing.

Look at picture **A**. It shows one cell. Notice the dark part inside the cell. That is where the cell will split in two.

Look at picture **B**. The dark part has just split in two. The two dark parts move away from each other. And a wall forms between them.

At a certain point, the wall is complete. When that happens, how many cells are there?

Right! There are two cells. In picture **C**, you can see what the two cells look like. Each cell is exactly like the other.

Suppose you cut your finger. Some skin cells are *destroyed*—killed. But in a few days, the cut heals. The skin over the cut is smooth. You have new skin cells. How does that happen?

Right! The cells around the cut divide. And they make new skin cells.

A

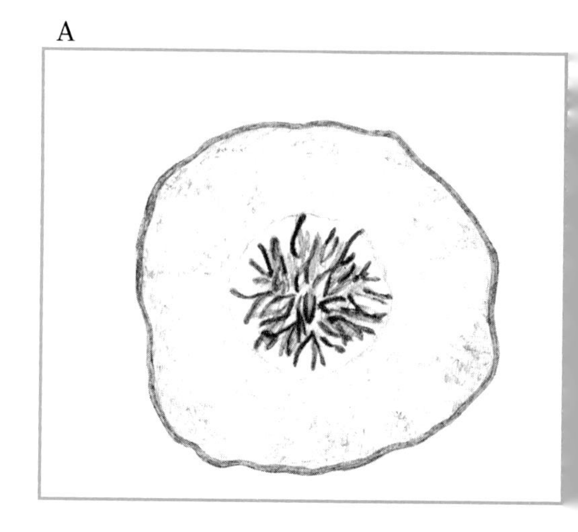

B

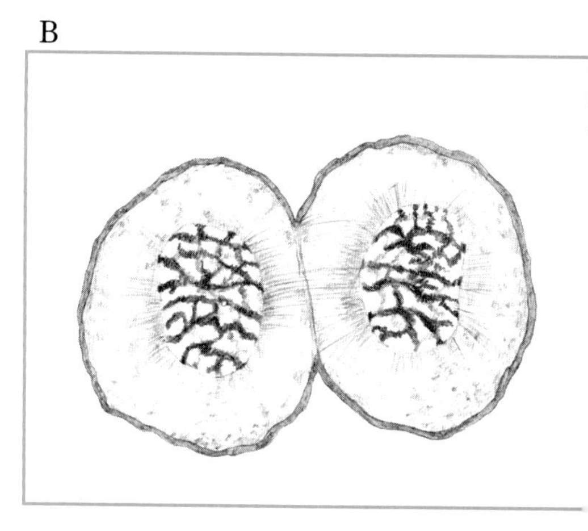

C

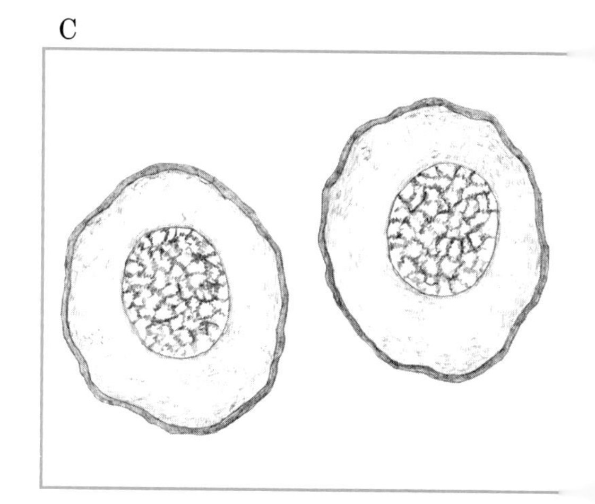

Close-Up of a Tissue

You can see what a tissue looks like. Get some raw hamburger meat. (Hamburger is muscle tissue. It looks a lot like muscle tissue that's in the human body.) Get these materials:

- One microscope
- One dropper with water
- One glass slide
- One cover slip

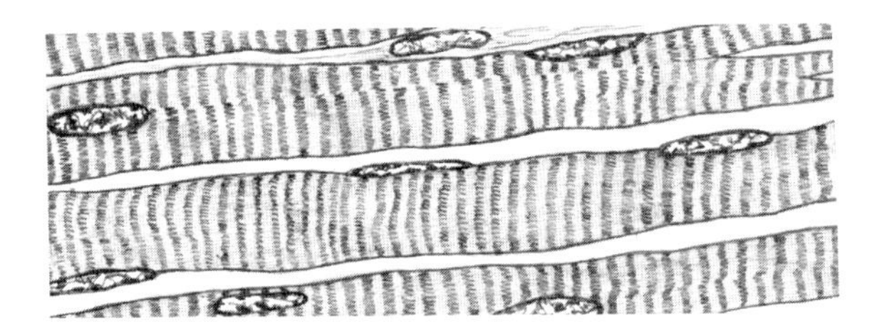

Muscle tissue

1 Put three drops of water on the slide. Add a little bit of hamburger to the water.

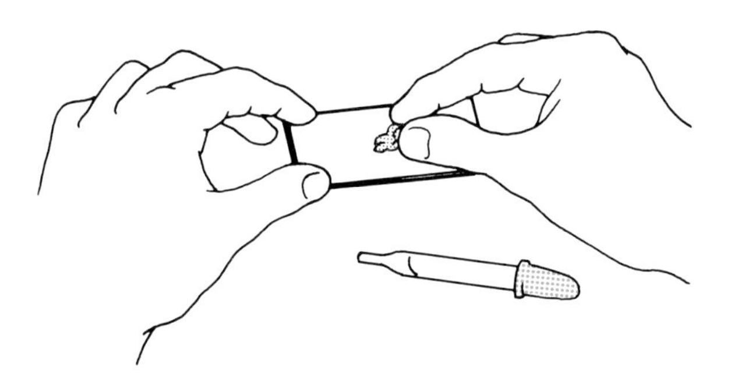

2 Put a cover slip over the hamburger. (Follow step 5 in "Close-Up of a Cell" on page 101.)

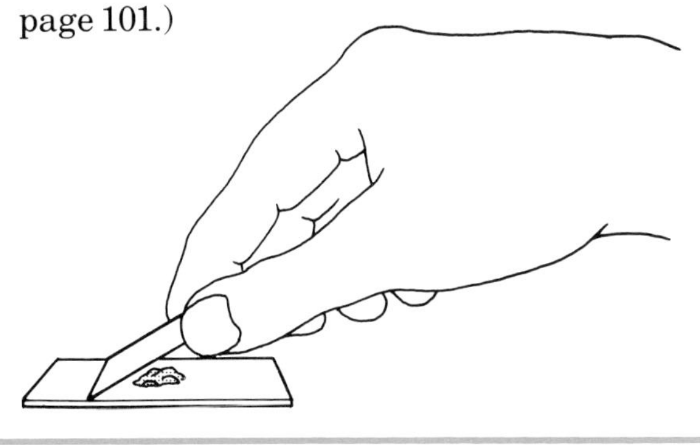

3 Put the slide in the microscope. Turn to low power. Focus the lens. You should see tissue—cells that look alike. Now turn to high power. On a separate piece of paper, draw the tissue.

Review

Show what you learned in this unit. Use the words in the list to answer the questions.

cell	tissue	organ
organ system	divide	

1. What is made up of organs that work together in your body to do a big job?
2. What do cells do when they split in half?
3. What is the smallest living thing?
4. What is made up of different kinds of tissues?
5. What is made up of the same kind of cells?

Check These Out

1. Make a Science Notebook for Human Systems. Divide your Science Notebook into three sections: *Science Work, Keep Fit,* and *In the News.*

   At the end of every unit, you'll find suggestions for activities you can do. Pick something for each section of your notebook. Put each thing you do in the right section.

   Here are some ideas to start your Science Notebook:

   a. *Science Work:*
      - With a microscope, look at slides of different cells. Draw the cells, label your drawings, and put them in your Science Notebook.
      - See what bone cells look like. Get a bone from a butcher. Dry the bone. Then carefully scrape it with a sharp blade. Make a slide with the scrapings and look at them under the microscope. Draw and label what you see. Put the drawing in your notebook.
      - A cell is made up of different parts. Look in the encyclopedia and find out what the parts are. Then make a drawing of a cell and label its different parts.
      - A cell lives for a certain number of days. Find out how long these cells live: the nerve cell, the red blood cell, and the skin cell. Take notes and put them in your Science Notebook. Then give (or tape) an oral report to the class about the cells.

   b. *Keep Fit:* Your cells need *minerals* and *vitamins*. Find out what they are. Put your notes in your Science Notebook.

   c. *In the News:* Sometimes sick cells in a tissue divide and outnumber the healthy cells. This is called *cancer.* In magazines or newspapers, find articles about cancer. Cut out those articles. Paste them on notepaper and put them in your Science Notebook.

Unit 2

Moving Parts

Do you like to dance? Run? Play basketball? When you do any of those things, your body moves in many different ways. Two organ systems work together so you can move.

- What organ system holds you up?
- What two organ systems let you move?

You'll learn the answers in this unit.

Before You Start

You'll be using the science words below. Find out what they mean. Look them up in the Glossary. On a separate piece of paper, write what the words mean.

1. **joint**
2. **skeleton**

Your Bones

Stand up. What keeps your body from falling?

Right! Your *skeleton* keeps your body from falling. It *supports* your body.

Your skeleton makes up your **skeletal system**. It is made up of about 206 bones. They protect the soft parts of your body and give your body its human shape.

Bones are organs. They are made up of tissues. The outside of a bone is hard tissue. Minerals such as calcium are stored there. The inside of a bone is soft tissue, or *marrow*. Red blood cells are made in the marrow of certain bones.

All the bones are *connected* to each other. They are joined together by tissues called *ligaments*. Another tissue—*cartilage*—is between the bones. The cartilage protects the ends of the bones. It keeps the bones from rubbing against each other.

Look at the diagram. Notice that the bones have different shapes. Some bones are called *flat bones*. They curve around soft organs such as your heart and lungs. They protect those organs. For example, your ribs protect your lungs and your heart.

Some bones are called *long bones*. They are round and straight. They let you move your body. The bones in your arms are long bones. What other parts of your body have long bones?

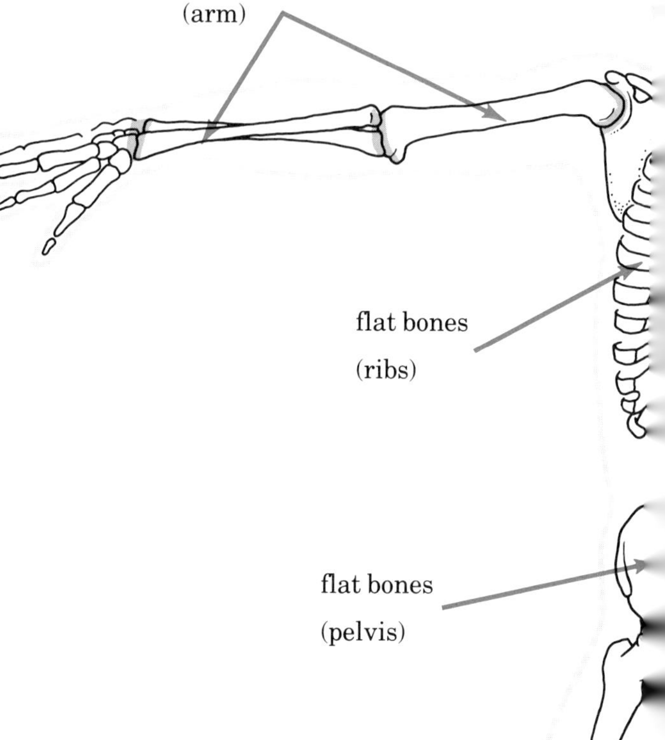

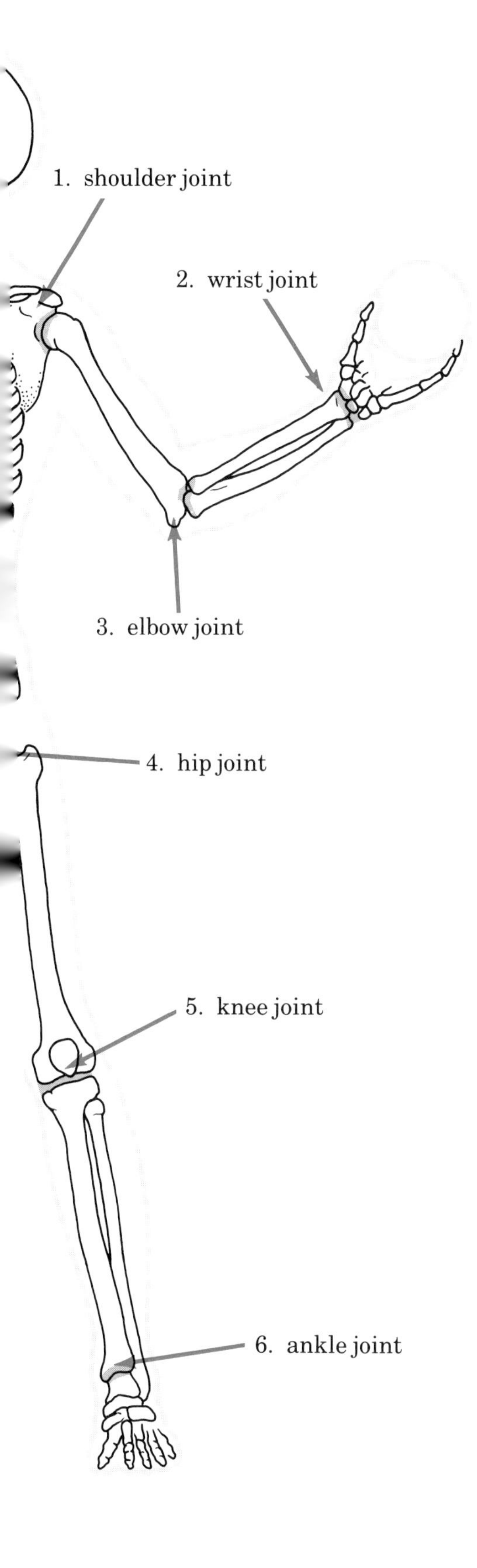

Your Joints

Stretch out your hand. Now bend your fingers. See how each finger is made up of several bones? You can bend your fingers because of certain parts at the ends of those bones. What are those parts called?

Right! Those parts are called *joints*.

Your skeleton has different kinds of joints that let you move. One of those is called a *hinge* joint. Think of the hinges on a door. Your hinge joint is something like a door hinge. It lets you move parts of your body the same way a door moves: back and forth.

The knuckles on your fingers are hinge joints. Your knee is a hinge joint on your leg. What hinge joint is on your arm?

Another kind of joint you have is called a *ball-and-socket* joint. It lets you move parts of your body in any direction. You can find that kind of joint in your shoulder, at the top of your arm. Touch that joint. Move your arm in wide circles. Then bend your arm. See how the two movements are different?

Now look at the diagram of the skeleton. The shoulder joint is a ball-and-socket joint. What kind of joint is in the wrist? Elbow? Hip? Knee? Ankle?

Muscles!

Blink your eyes. Swallow. Stand up and take a step. Every time you do those things, parts of your body move.

What parts move when you blink your eyes?

What parts move when you swallow?

What parts move when you take a step?

Your body moves in those ways because of a certain kind of organ: *muscles*. All the muscles in your body make up your **muscular system**.

Your body has three kinds of muscles. One kind of muscle moves just your skeleton. Those muscles are called *skeletal muscles*.

Another kind moves parts of organs inside your body. Those muscles are called *smooth muscles*. Your stomach has smooth muscles.

The third kind of muscle is very special. It makes up a whole organ. That organ pumps blood through your body. What is that organ?

Right! It's your heart.

Now look at the diagram. It shows the muscles that move your body. Which kind are they: heart muscle, smooth muscle, or skeletal muscle?

Find the two muscles that are on the arm. What are their names?

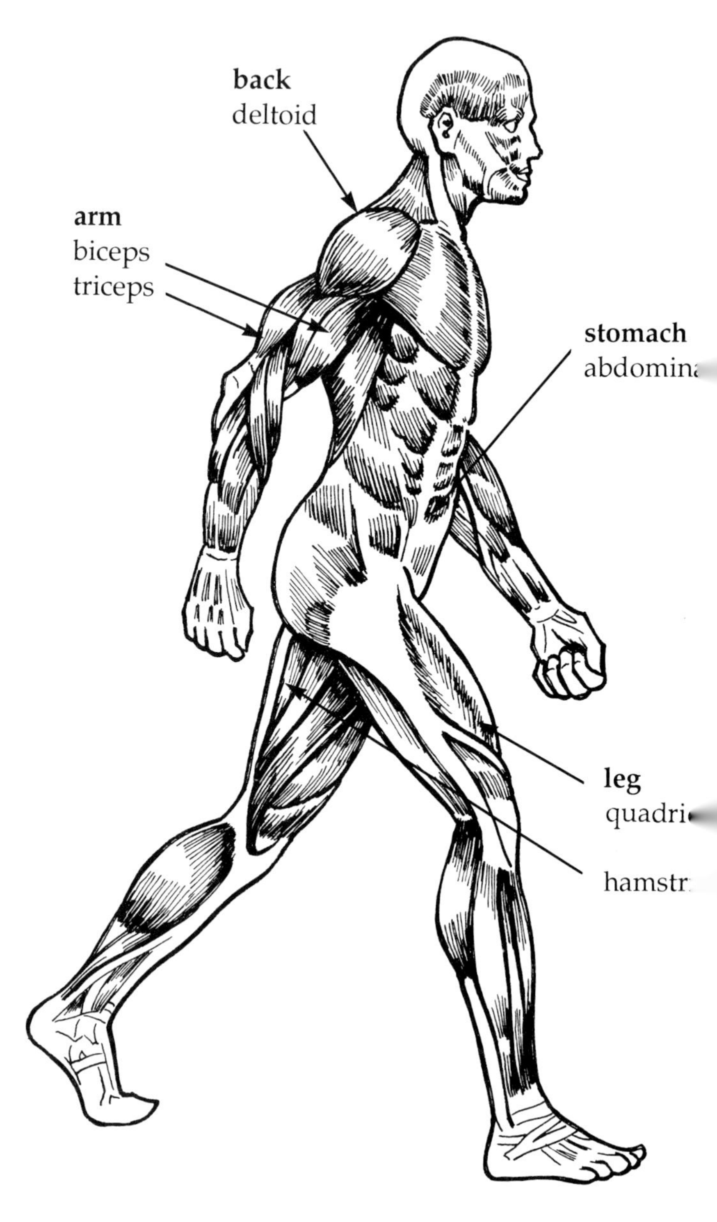

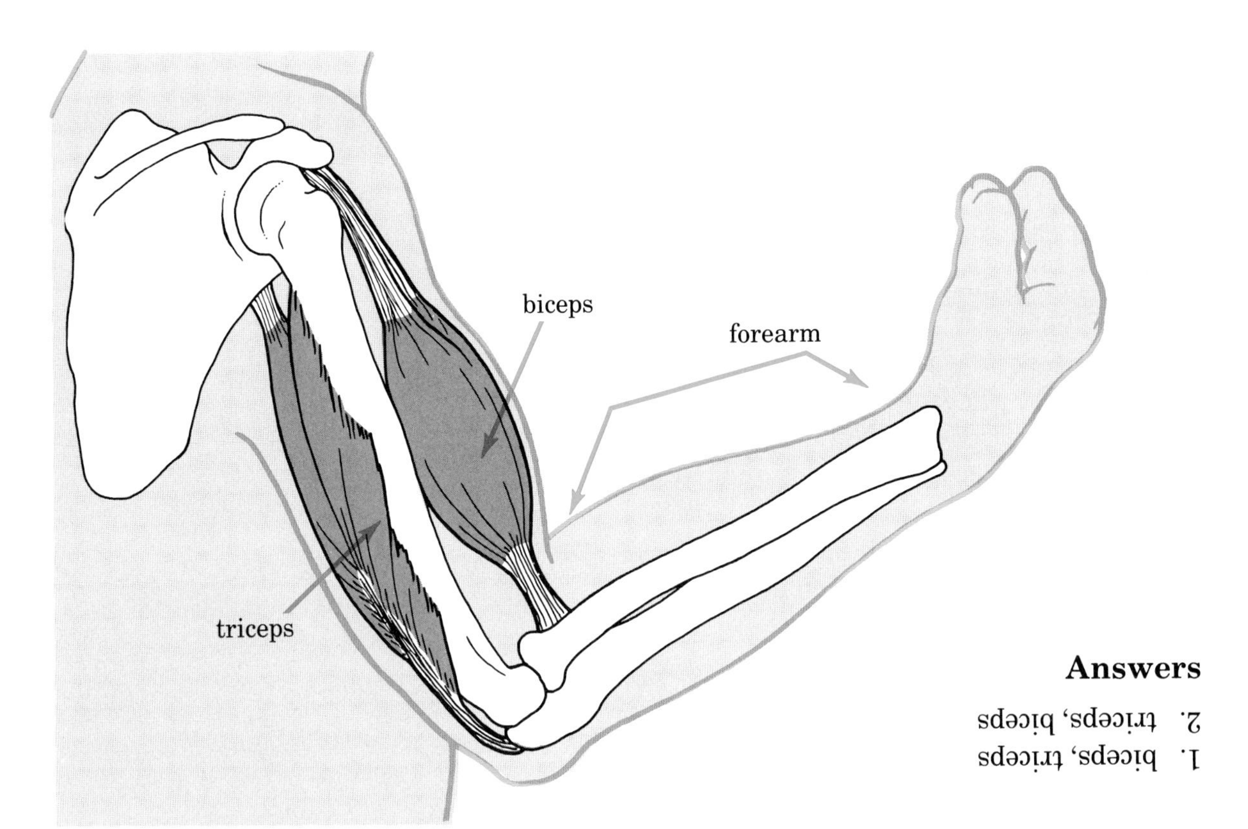

Answers

1. biceps, triceps
2. triceps, biceps

How Your Muscles Work

Put your elbow on your desk. Move your forearm up and down. (That's the part of your arm below your elbow.) How many muscles do you use?

You use two skeletal muscles. You use one muscle to lift your forearm. You use another to lower it.

Skeletal muscles are made of long tissues. The ends of those tissues are attached to bones. You move when the muscles pull those bones.

A muscle can pull a bone in only one direction. So two muscles are needed to move a body part. One muscle pulls a bone in one direction. Another muscle pulls it back.

When a muscle pulls a bone, the muscle **contracts** and gets shorter. It pulls the bone toward it. The muscle **relaxes** and gets longer when another muscle pulls the bone.

Your biceps and triceps are two muscles that work together. They move your forearm. Put your hand on the middle of your upper arm. Bend your arm. You should feel a lump. That is your biceps. It pulls your forearm *up*. Does the biceps contract or relax?

Right! Your biceps contracts.

Now lower your forearm. You should feel the biceps relax. That means your triceps is contracting. The triceps pulls the bone in your forearm *down*.

Look at the diagram. It shows the muscles that move your forearm. Answer the questions with *biceps* or *triceps*. (The answers are upside down.)

1. When the forearm is raised, which muscle contracts? Which muscle relaxes?
2. When the forearm is lowered, which muscle contracts? Which muscle relaxes?

Review

Use the words you learned in this unit to answer the questions. The page where you learned the word is listed after each question.

1. What does a muscle do when it gets shorter? (page 109)
2. What supports your body? (page 106)
3. What is the name of the system that makes the parts of your body move? (page 108)
4. What kind of organ are biceps and triceps? (page 108)
5. What makes up your skeletal system? (page 106)
6. What does a muscle do when it gets longer? (page 109)
7. What is the name for the places where ends of bones meet? (page 107)

Check These Out

1. For your Science Notebook:
   a. *Science Work:*
      - Get prepared slides of skeletal muscle and smooth muscle. Look at each one in the microscope. Draw what you see.
      - Get a long bone of an animal from a butcher. Find the different parts of the bone. (Use an encyclopedia or science book to help you.) Make a drawing and label the parts.
      - Get a whole leg of a chicken. (The leg bone and thigh bone are connected.) Find these tissues: cartilage, marrow, and bony tissue.

   b. *Keep Fit:* Good posture keeps all your organ systems working in good order. Find out what good posture looks like. Then find out what kind of posture you have.

   c. *In the News:* Many diseases can affect your skeletal and muscular systems. For example, arthritis is a disease in the joints. Find an article about a disease that affects the skeleton or muscles.
2. As you work through this section, you may want to find out more about the human body. You can find out more by looking in an encyclopedia or by getting books from a library. You can also talk to an expert, such as a doctor or a science teacher.

   Here are some things you may want to find out:
   - What is muscle tone?
   - What is aerobic exercise?
   - What is the Achilles tendon? The funny bone?

Unit 3

Staying Alive

Imagine that you're hiking up a mountain. You're tired. You're out of breath, thirsty, and hungry.

Finally, you stop to rest. You drink water. You take big breaths of fresh mountain air. You feel great! Then you quickly get out your lunch.

Your body needs the things that are in air, water, and food. Without them you would die. Two organ systems get those things for your body.

- What are the things your body needs?
- What organ systems get those things?
- How does each organ system work?

You'll learn the answers in this unit.

Before You Start

You'll be using the science words below. Find out what they mean. Look them up in the Glossary. On a separate piece of paper, write what the words mean.

1. **digest**
2. **enzyme**
3. **waste products**

Your Respiratory System

Every cell in your body needs **oxygen**. Oxygen is one of the gases in the air. How does oxygen get into your body?

Right! Your body gets oxygen when you breathe.

The organs that let you breathe make up your **respiratory system**. That system takes the oxygen out of the air that you breathe *in*. It also gets rid of a waste product that cells make when they use oxygen. That waste product is *carbon dioxide*.

How do you think your body gets rid of carbon dioxide?

Right! Each time you breathe *out*, your body gets rid of some carbon dioxide.

Look at the diagram on the next page. It shows the organs that make up your respiratory system. Your nose, mouth, and windpipe are part of that system. What two large organs are also in that system?

Yes! Your two lungs are in that system.

Your lungs are next to a muscle that helps you breathe. That muscle is attached to your ribs. When the muscle contracts, you breathe in. And when it relaxes, you breathe out. What is that muscle's name? (The answer is upside down.)

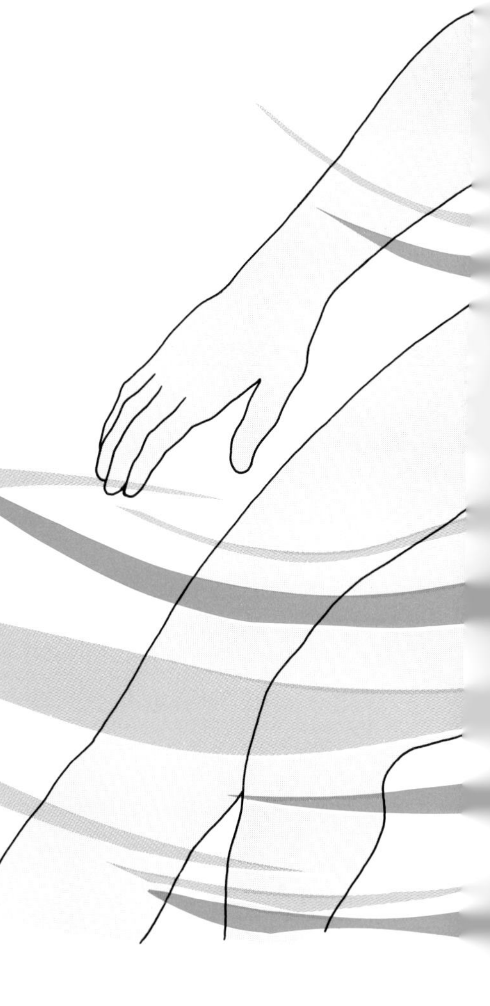

Answer

the diaphragm

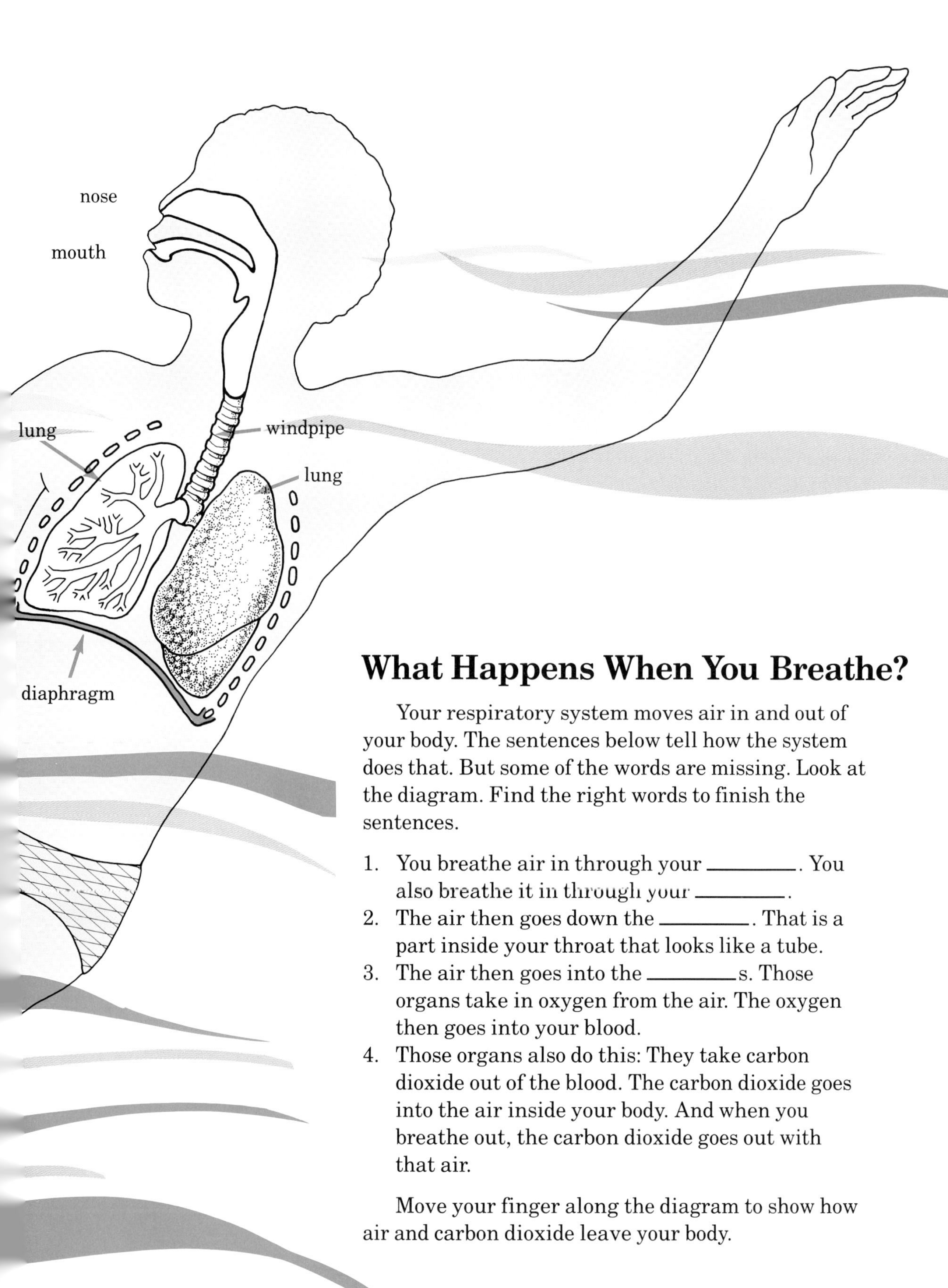

What Happens When You Breathe?

Your respiratory system moves air in and out of your body. The sentences below tell how the system does that. But some of the words are missing. Look at the diagram. Find the right words to finish the sentences.

1. You breathe air in through your ________. You also breathe it in through your ________.
2. The air then goes down the ________. That is a part inside your throat that looks like a tube.
3. The air then goes into the ________s. Those organs take in oxygen from the air. The oxygen then goes into your blood.
4. Those organs also do this: They take carbon dioxide out of the blood. The carbon dioxide goes into the air inside your body. And when you breathe out, the carbon dioxide goes out with that air.

Move your finger along the diagram to show how air and carbon dioxide leave your body.

Your Digestive System

How do you feel when you don't eat for a while? Tired? Or full of energy?

Your body gets tired when you don't eat for a while. That's because food gives your body energy. It also gives your body **nutrients**—the things cells need to stay alive.

Before your body can use what you eat, it must *digest* that food: It must break food down into very tiny bits. The organs that digest food make up your **digestive system**.

The diagram on this page shows the digestive system. The system looks like a long tube that goes through the body. Food that you eat travels through that tube. Look at the diagram. What organ first starts to digest the food?

Right! The mouth begins digesting food. Many foods are hard and solid. Your system breaks those foods down in several ways. One way is this: It cuts food into small pieces. Your teeth do that when you chew.

Your system also breaks food down by mashing it. That's what happens to food in your stomach.

Strong chemicals can also break food down. Your digestive system has glands that make *enzymes* and *digestive juices*. Those chemicals can break food down into a liquid form.

You can see how enzymes work. The saliva in your mouth has an enzyme. Put a cracker in your mouth. Don't chew it. What happens to the cracker?

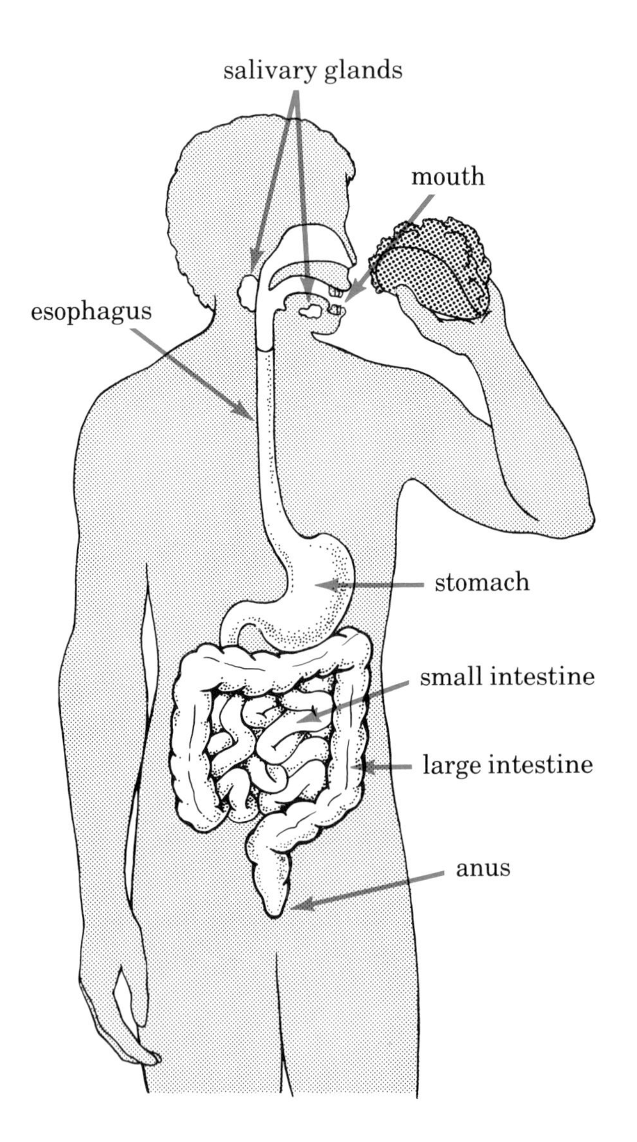

What Happens to the Food You Eat?

The sentences below tell what happens to food you eat. Read the sentences. Then read the questions after them. Look at the diagram on the other page and answer the questions.

1. You put food into your mouth and chew it. Your teeth cut the food into small pieces. Glands in your mouth give off an enzyme in saliva. It begins to digest the food. What are the glands?
2. You swallow the pieces of food. They move down a part that is like a tube. What is that part?
3. The pieces of food next go into an organ. It mashes the pieces into tiny bits. What is that organ?
4. The tiny bits of food then go into another organ. There, enzymes break the bits down into nutrients. (The nutrients then go into the blood.) What is that organ?
5. Some parts of food can't be broken down. Those parts move all through the digestive system. They finally go into an organ and become solid wastes. What is that organ?
6. The solid wastes are pushed out of the body. Through what part do the wastes go out?

A Model of the Stomach

Your stomach makes an acid that helps it digest food. That acid is called *hydrochloric acid*. You can easily see how that acid breaks down food. You'll need these materials:

- One small jar (such as a pickle jar)
- One small piece of cooked beef
- Some dilute hydrochloric acid

1 Put the piece of beef in the jar.

---

2 Ask your teacher to add some dilute hydrochloric acid. (That acid is like the acid in your stomach.) It should cover the beef.

CAUTION: Some acids burn! So don't touch acids. If an acid spills on you, wash it off right away.

---

3 Put the jar in a safe place for one day. On the next day, look at the beef. Then answer these questions.

a. What did the beef look like before it was covered with acid?
b. What did the beef look like one day later?
c. What happened to the beef?

Review

Show what you learned in this unit. Answer the questions. The words you'll need are listed below.

digestive system	respiratory system
enzymes	nutrients
oxygen	carbon dioxide

1. What is the system that moves air in and out of your body?
2. What do your cells use that's in the air you breathe?
3. What is the waste product that your cells make from the air you breathe?
4. What is the system that helps your body use food?
5. What are the parts of food that your cells need?
6. What are the chemicals that help you digest food?

Check These Out

1. For your Science Notebook:
   a. *Science Work:*
      - Carbohydrates are nutrients that your body needs. Find out what foods have those nutrients. Get a bottle of iodine, bread, milk, a fruit, and a piece of meat. Put drops of iodine on each food. If the food has a carbohydrate, the iodine will turn a dark blue. Which foods have carbohydrates?
      - Get a prepared frog. With your class, dissect it (cut it open). Draw its digestive and respiratory systems.

   b. *Keep Fit:* Carbohydrates, proteins, and fats are the nutrients you need. Find out what each nutrient does for you. What foods supply each nutrient?

   c. *In the News:* Many people go on *fad diets* to lose weight. Find an article about a fad diet.
2. Smoking cigarettes can cause lung cancer. Invite a doctor or a nurse to talk about lung cancer.
3. Invite a dentist to speak to your class.
4. The liver, pancreas, and gall bladder help digest food. Find a picture that shows those organs. (Look in an encyclopedia.) What else do they do?
5. Here are more things you may want to find out:
   - Who was Dr. William Beaumont?
   - What is anorexia nervosa? Bulimia?
   - What is *metabolism?*

Unit 4

The Movers and Removers

Your cells need oxygen and nutrients. Those things must somehow move all through your body. They must get to every one of your billions of cells.

Your cells make waste products when they use oxygen and nutrients. Those waste products must move out of your body.

- What organ systems move oxygen and nutrients?
- What organ systems move waste products?
- Why must the body get rid of waste products?

You'll learn the answers in this unit.

Before You Start

You'll be using the science words below. Find out what they mean. Look them up in the Glossary. On a separate piece of paper, write what the words mean.

1. **excrete**
2. **lymph**

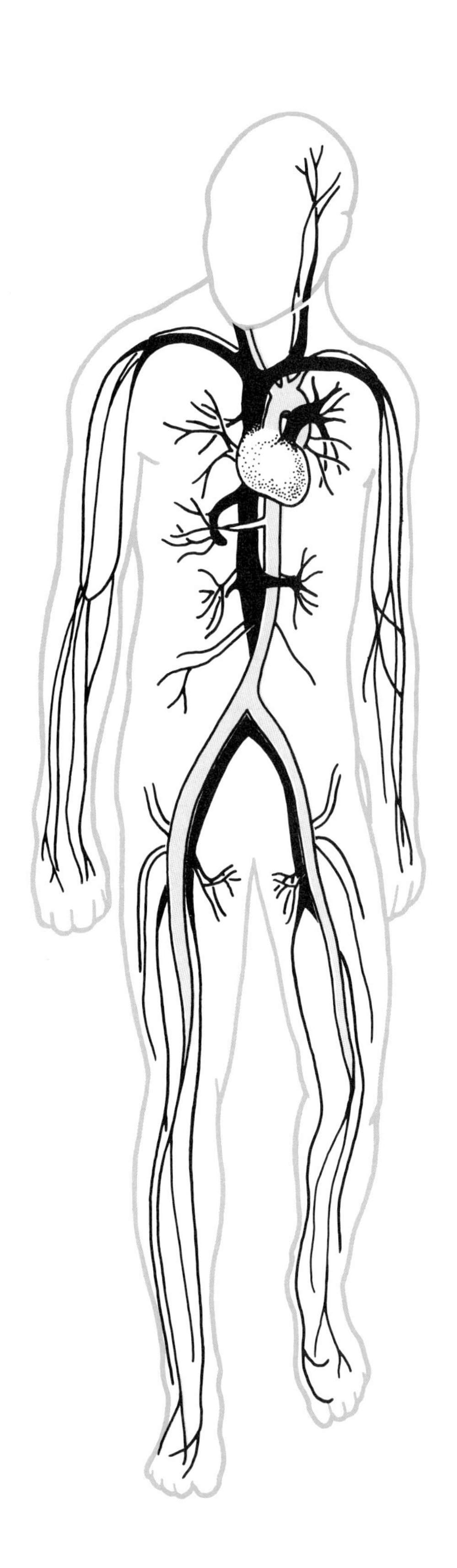

The Stream of Life

You have something that moves like a river inside your body. It flows to every part of your body. As it flows it carries things. What do you think it is?

Did you say *blood?* Right! Blood moves throughout your body. It flows through parts called *blood vessels*.

Blood and blood vessels are part of your **circulatory system**. That system moves things such as oxygen and nutrients throughout your body. What important organ is also part of that system? Look at the diagram.

Yes! It's your heart.

Blood always moves one way: It flows from your heart to your cells, then back to your heart, and so on.

Blood flows away from your heart through blood vessels called *arteries*. That blood carries oxygen from the lungs. It carries nutrients from your small intestine. It carries those things to the cells. As the blood gets closer to the cells, it flows through smaller and smaller arteries.

In the tissues, the blood flows through very thin and tiny blood vessels called *capillaries*. The oxygen and nutrients in the blood leave the capillaries and go to the cells. Waste products leave the cells and go to the capillaries.

The blood in the capillaries now carries waste products. It flows to bigger blood vessels called *veins*. In the veins, the blood now flows toward an organ. What organ is it?

Make a Model of a Heart

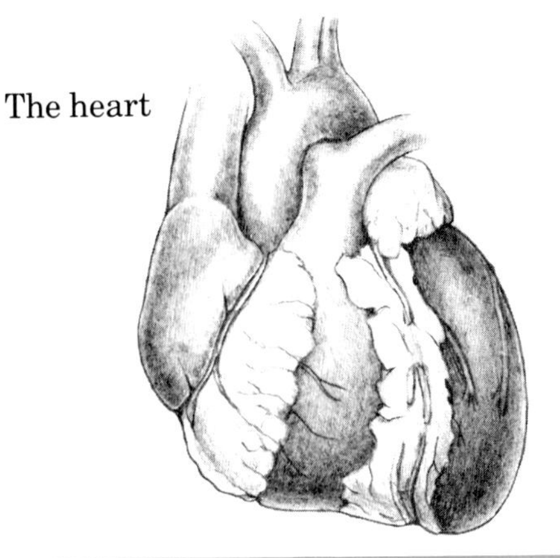
The heart

Blood *circulates*—moves—through your body because of your heart. Your heart is something like a bag made of muscle. When the muscle relaxes, blood from the veins goes into the heart. When the muscle contracts, it squeezes the blood out of the heart. That blood is pushed into the arteries.

Make a model of the heart, and see how that happens. You will need these materials:

- ½ cup of flour
- ½ cup of water
- One small, heavy plastic bag
- One clear plastic straw
- Tape

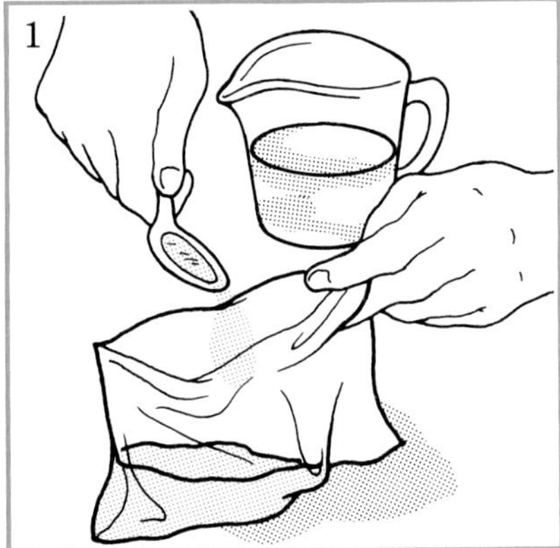

1 Mix ½ cup of flour and ½ cup of water. Pour the mixture into the plastic bag. The mixture stands for blood.

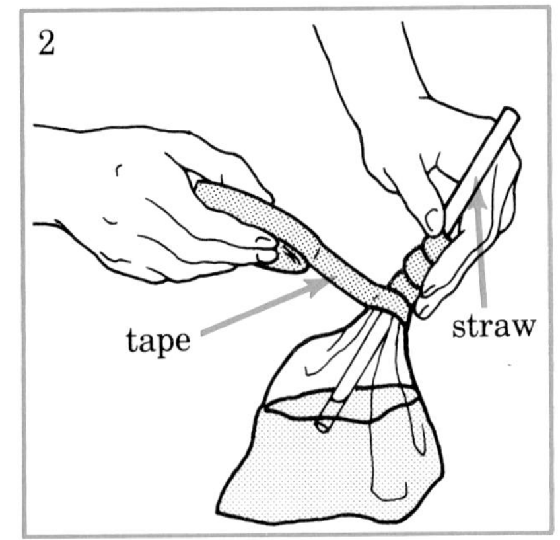

2 Put the straw into the bag. The straw stands for the artery. Gather the top of the bag around the straw. Wrap tape tightly around it. Make sure the tape covers the top completely.

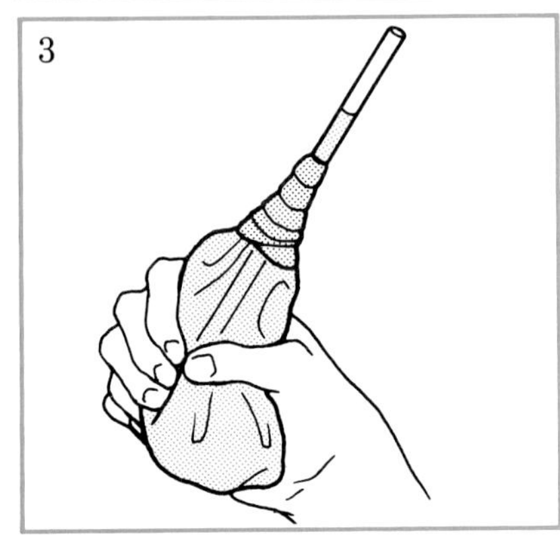

3 Wrap your hand around the bag. Your hand and the bag stand for the heart. Gently squeeze the bag, open your hand, and squeeze gently again. Do that for a while. Where does the mixture go?

The mixture goes into the straw. That's the way blood leaves the heart and goes into the arteries. Each time you squeeze the bag, more mixture is pushed into the straw. It pushes the mixture that is already there. That's the way blood is pushed through the arteries.

Listen to your heart. Hear it beat? Every time it beats, it contracts, and that pushes blood out of the heart and through the blood vessels.

What's happening in your heart when you don't hear the beat?

Your Lymphatic System

Another organ system moves things throughout your body. It is your **lymphatic system**. It is made up of parts called *lymph, lymphatic vessels*, and *lymph nodes*.

Lymph is one of the fluids that flow in your *bloodstream*. When your blood flows through the capillaries, the lymph leaks out. It flows through the tissues. As it flows, it carries things such as nutrients to the cells. It also carries things away from the cells. What do you think it carries away?

Yes! It carries away waste products that cells make. It also carries away germs and bacteria that have gotten into the tissue.

The lymph returns to the bloodstream. It returns through lymphatic vessels. (Those vessels cover every part of the body.)

As the lymph moves through the vessels, it passes through many lymph nodes. Those are tiny, round glands. They trap and kill the germs and bacteria that are in the lymph.

You can sometimes feel your lymph nodes. When you have an infection, they sometimes swell up. That's because they are full of the germs and bacteria they have trapped.

The diagram shows some lymphatic vessels. It also shows the main lymph nodes. On your own body, find the places where those lymph nodes are. What are those places?

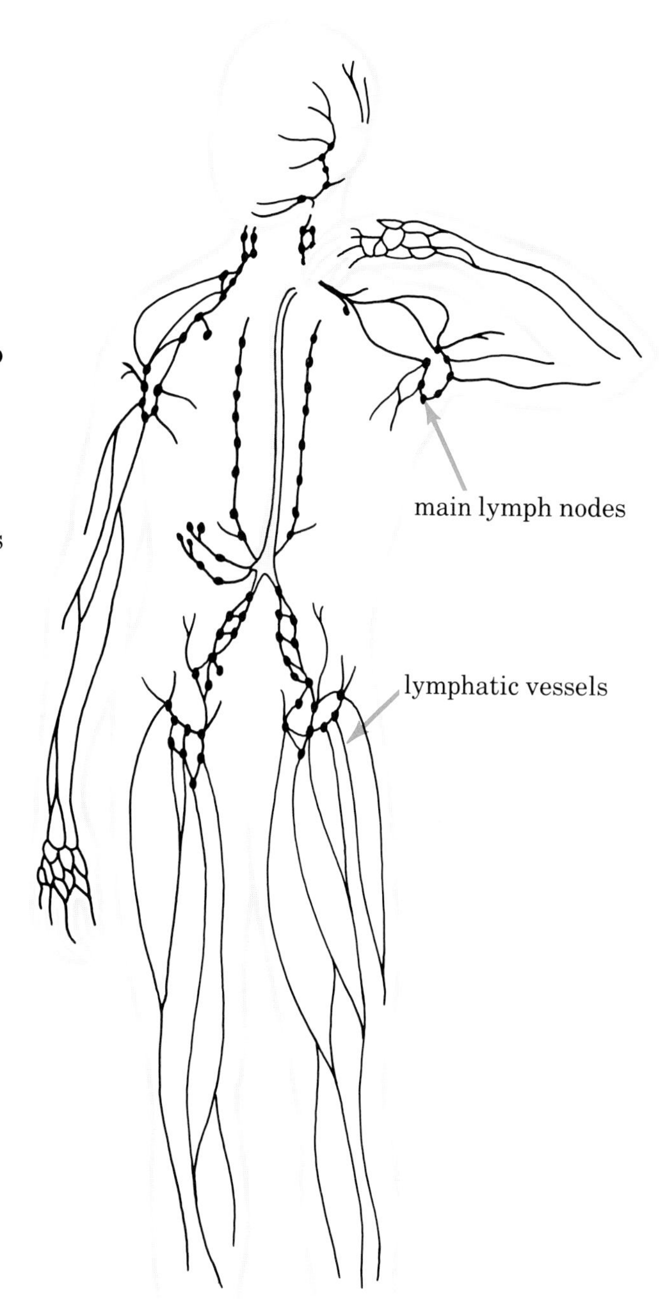

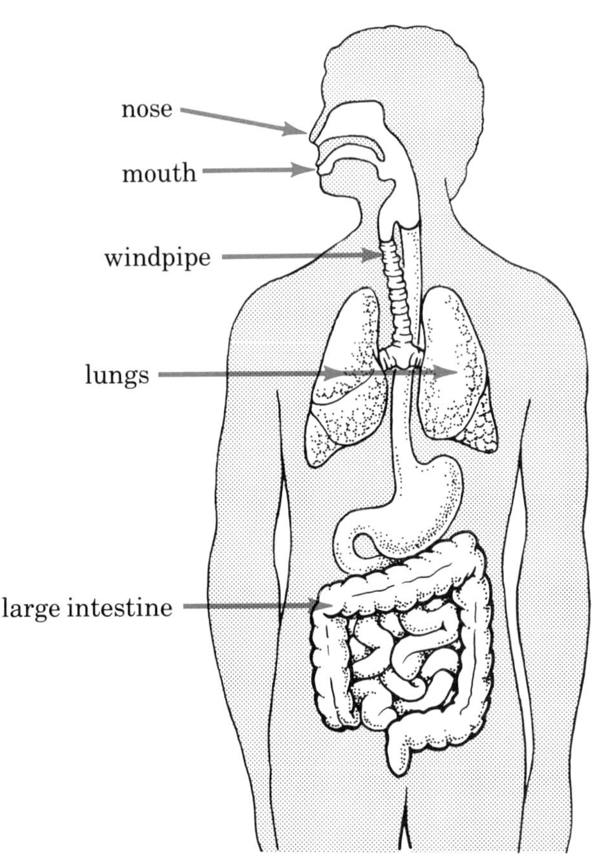

Getting Rid of Wastes

Waste products are always forming in your body. You learned about one waste product that goes out when you breathe. What is it?

Right! That waste product is *carbon dioxide*.

Carbon dioxide forms in your cells. *Salts, chemicals,* and *water* are also formed in the cells. They are all waste products that form when cells use oxygen and nutrients.

Foods that can't be digested also become waste products. That kind of waste forms in the large intestine of your digestive system.

Your body must get rid of all the waste products that form in it. What do you think would happen if waste products stayed in the body?

Waste products are like poisons. If they stay in your body, the tissues become sick and die.

Several organs and organ systems work together to excrete waste products. Your digestive system excretes the solid wastes that form in your large intestine. Your respiratory system excretes carbon dioxide that the cells make. Your **excretory system**—another organ system—excretes the other waste products that the cells make. (You'll learn more about this system on page 124.)

The diagram on this page shows some organs in the digestive and respiratory systems. Those organs help get waste products out of the body. Say the names of those organs to yourself.

A Carbon Dioxide Test

Your respiratory system excretes carbon dioxide. You can't see or smell carbon dioxide. But there's a way to show that your body excretes this waste product.

You'll need these things:

- A chemical called *bromthymol blue*
- One small jar
- One straw

1 Pour bromothymol blue into the jar. Fill it one-fourth full.

2 Put a straw in the jar. Gently *blow* into the straw for 30 seconds.

What's Happening?

1. What color is the chemical before you blow into the jar?
2. What color is the chemical after you blow into the jar?
3. What changes the color of the chemical?
   Right! The carbon dioxide in the air that you breathe out changes the color.
4. If oxygen is mixed with the chemical, the color will change. What color will the chemical be?
   Right! It will turn blue again. (Try it!)

Your Excretory System

Carbon dioxide is one waste product that your cells make. What's another waste product? (Look back on page 122 for help.)

Salts, chemicals, and water are the other waste products.

You excrete some of those waste products through your skin. That happens when you sweat. (Notice the salt on your skin after the sweat dries up.)

But most of those waste products leave your body through the *excretory system*. Look at the diagram. It shows the main organs in that system. Find the kidneys.

Blood is always flowing through the kidneys. As it flows through, the kidneys take the waste products out of the blood.

The waste products then go into the two small tubes. Where do they go next? (Look at the diagram.)

Right! The waste products pass into the *bladder*.

The bladder holds the waste until the body excretes it. Then the waste passes out of the body through another small tube. The waste is called *urine*. What is the small tube called?

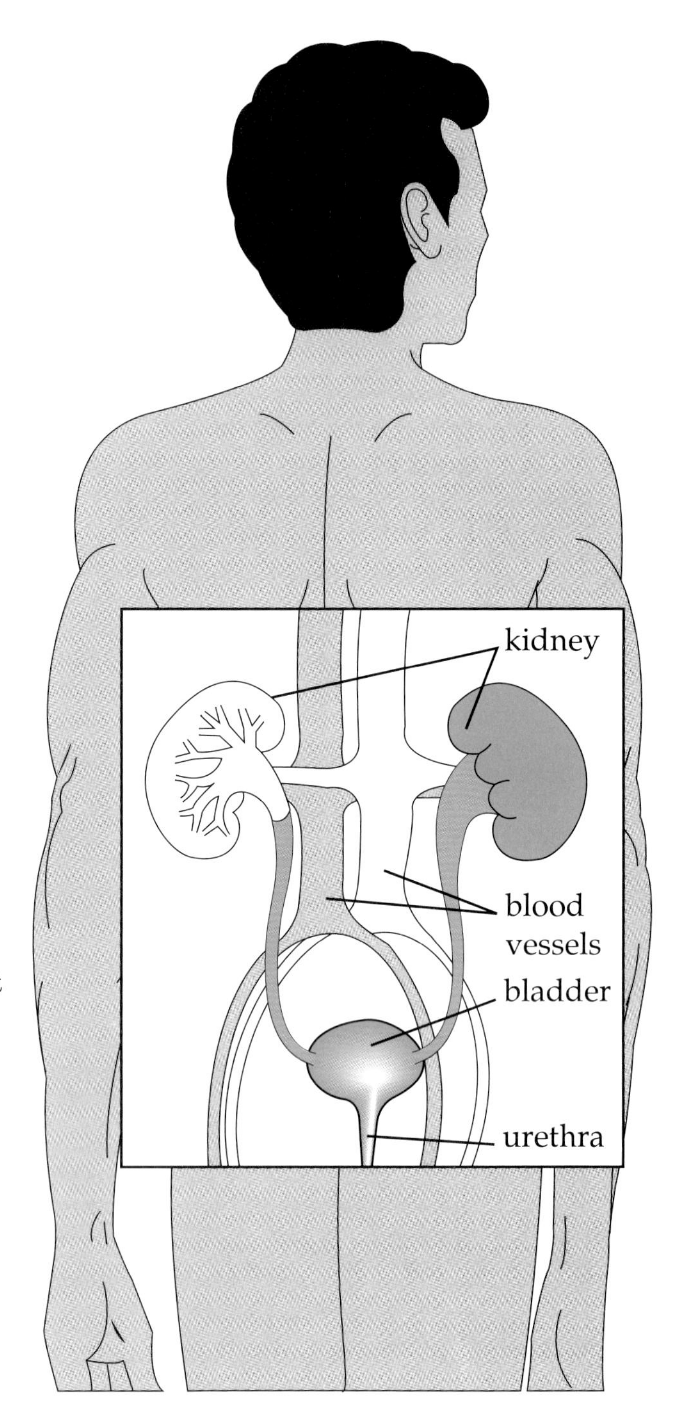

Review

Show what you learned in this unit. Finish the sentences. Match the first part of the sentence on the left with the correct words on the right.

1. Your circulatory system carries oxygen,
2. Your lymphatic system carries
3. Your digestive system excretes solid
4. Your respiratory system takes out carbon
5. Your excretory system excretes salts,

a. waste products that form in the large intestine.
b. nutrients, and waste products through the body.
c. dioxide that forms in cells.
d. germs and bacteria to the lymph nodes.
e. chemicals, and water that form in the cells.

Check These Out

1. For your Science Notebook:
   a. *Science Work:*
      - Get an animal heart from a meat shop. Cut the heart in half, and draw what you see.
      - Your heart pumps about five quarts of blood in one minute. How many gallons does it pump in 24 hours? (There are four quarts in a gallon.)
      - White blood cells protect your body from infections. Find out how those cells do that. Make a report to your class.
      - What happens to your body when you have a cold? Find out. Then write a report about it.

   b. *Keep Fit:*
      - How does *aerobic* exercise keep your respiratory and circulatory systems healthy and fit? Find out.
      - Fiber is needed for a good diet. Find out what fiber does. Also find out what foods have fiber.

   c. *In the News:* People can receive a "new" heart. It may be an artificial heart or a heart from another person (a heart transplant). Find an article about someone who received a "new" heart.
2. Your local hospital probably has a *blood drive* every year. Find out what it is. Then visit the next blood drive your hospital holds.
3. Here are more things you may want to find out:
   - What is leukemia? Sickle cell anemia?
   - What does a doctor check for in a physical exam?
   - What are vaccines?
   - What is a cardiac arrest?

Unit 5

Systems That Control Your Body

Your skeletal and muscular systems let your body move. Your respiratory and digestive systems get oxygen and nutrients for your body. Your circulatory, lymphatic, and excretory systems move materials through your body.

All these systems work because of two other organ systems. Those systems control your body.

- What two organ systems control your body?
- What does each system do?

You'll learn the answers in this unit.

Before You Start

You'll be using the science words below. Find out what they mean. Look them up in the Glossary. On a separate piece of paper, write what the words mean.

1. **action**
2. **hormone**
3. **messages**

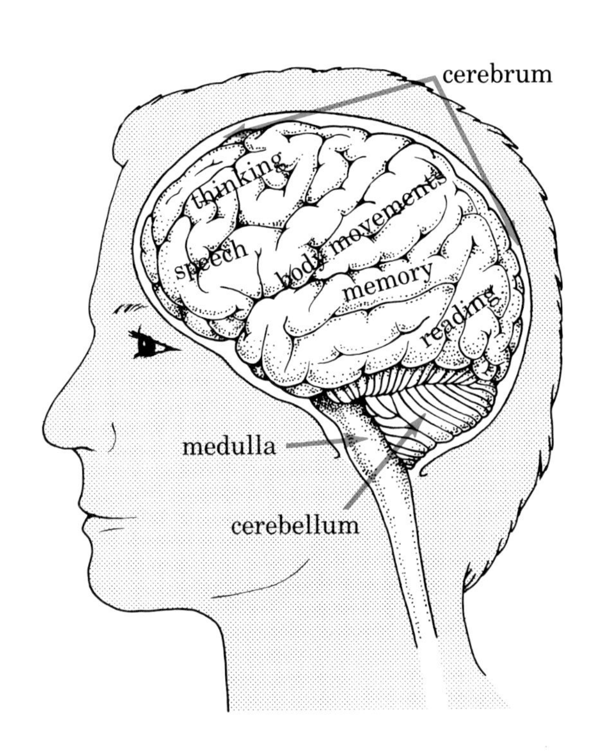

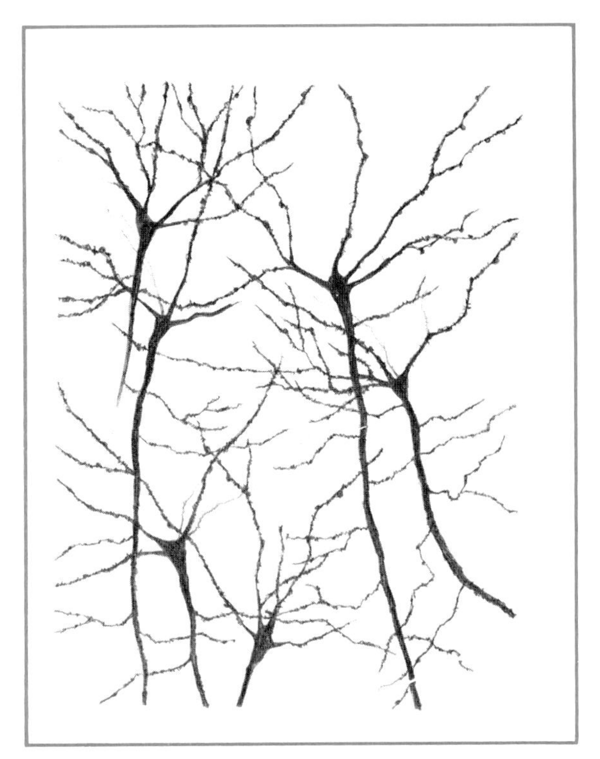

Brain nerve cells

Your Brain

Think about all the ways your body moves: It bends, stretches, twists, and so on. You have an organ that tells your body how to move. What do you think it is?

That's right! It's your *brain*.

The brain, *nerves*, and *sense organs* make up an organ system that controls every part of your body. That system is your **nervous system**. Here's how that system works:

Your brain sends messages to different parts of your body. The messages travel along special tissues called nerves. Those tissues are made up of nerve cells that are connected to each other. And the nerve cells form a path that goes from the brain to different parts of your body.

The brain is also made up of nerve cells. The picture above shows what some brain nerve cells look like. How are they different from your cheek cells? (Look at page 101.)

Different parts of your brain control the different things you do. Look at the diagram of the brain. Find the part of the brain that's called the *cerebrum*.

A part of the cerebrum controls your speech—the way you talk. Another part of the cerebrum controls your body movements. It controls the different parts of your body that move, such as your hands. It tells those body parts when and how to move. What other body parts are controlled by this part of the brain?

Your Sense Organs

You learned that your brain sends messages to every part of your body. Those messages tell the different body parts what to do.

Your brain also receives messages. What kind of messages do you think your brain receives?

Your brain receives facts about the world around you. Those facts tell the brain what things look like, feel like, and so on.

Certain organs get the information the brain needs. Those organs are the *sense organs*. They send messages to the brain. The brain then tells the body parts what to do. Your brain and sense organs are always sending messages back and forth. And it all happens in less than a second!

Several sense organs in your body gather information about the world. Your main sense organs are your nose, ears, eyes, tongue, and skin.

Each of those sense organs gathers certain information. For example, your nose gathers facts about the way things smell. What kind of information do your eyes gather? Your ears? Your tongue? Your skin?

Kinds of Action

Your brain controls every action in your body. Some actions, such as sitting or reading, are *voluntary* actions. That means you do them because you want to. When you want to sit, this happens: Your brain tells muscles in your body to move certain parts so you can sit.

Other actions, such as your blood moving through your body and your eyes blinking, are *involuntary.* That means your body does them automatically. You don't have to think about doing them.

Some actions that you do are listed below. Which ones are voluntary? Which ones are involuntary? (The answers are upside down.)

1. Walking
2. Breathing
3. Eating
4. Talking
5. Heart beating
6. Throwing a ball
7. Sweating

Look at the picture. What voluntary actions is that person doing?

What involuntary actions do you think that person's body is doing?

Answers

1. voluntary
2. involuntary
3. voluntary
4. voluntary
5. involuntary
6. voluntary
7. involuntary

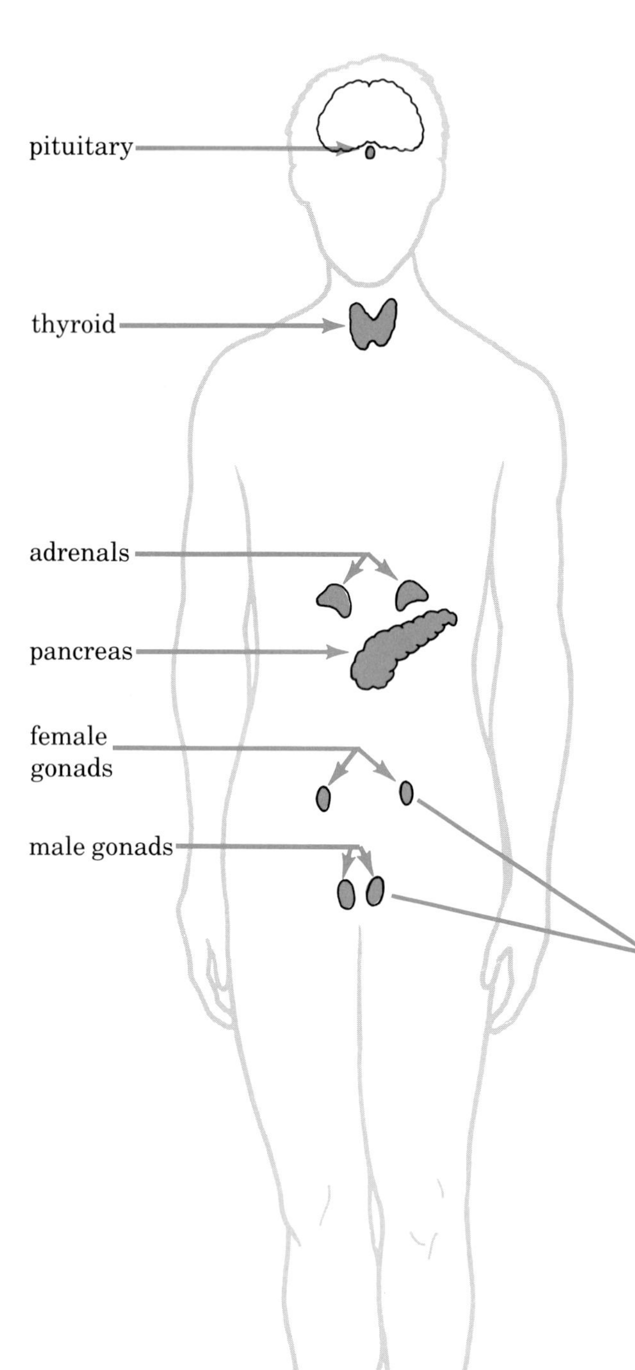

Your Endocrine System

Another organ system also controls your body. That system is your **endocrine system**.

The endocrine system is made up of *endocrine glands*. Those glands are scattered throughout your body. They make certain chemicals called *hormones*. Hormones make your cells do certain things, such as use nutrients quickly or slowly.

The diagram shows some important endocrine glands. Read the sentences below about each gland. Then find each gland on the diagram.

- The *pituitary gland* is behind the middle of the forehead. That gland makes hormones that control many of the other endocrine glands.
- The *thyroid gland* is in the neck. It makes hormones that control how cells change nutrients into energy.
- The *adrenal glands* are above the kidneys. They make *adrenaline*. Adrenaline gives you quick strength and energy.
- Part of the *pancreas* is an endocrine gland. It makes a hormone that controls how much sugar is turned into energy.
- The *gonads* control the organ system that makes new humans. They also control the changes that males and females go through during their teenage years.

The hormones are carried to cells by another organ system. What is it?

Right! Your circulatory system. The hormones from the glands go into your blood. Your blood then carries the hormones to the cells.

Glands in Action

What do your endocrine glands do? Answer these questions. Then check your answers. (The answers are upside down.)

1. Suppose you're 16 years old. Your body is changing. Your voice is changing. Hair is growing on your body. What endocrine glands are at work?
2. You're walking home. You see a friend about to cross the street. You notice a speeding car coming right at your friend. You run as fast as you can to save him. What glands send hormones into your body so you can run fast?
3. You're tired. You eat an orange, and you have energy again. The sugar in the orange gives your cells energy. One of your endocrine glands causes that to happen. What gland is it?
4. The food you ate for breakfast is digested. The nutrients are in your bloodstream. A gland sends out hormones that tell your cells to change the nutrients into energy. What gland is it?
5. Your endocrine glands are always making hormones. They do that because one gland controls them. What gland is it?

Answers

1. gonads or pituitary gland
2. adrenal glands
3. pancreas
4. thyroid gland
5. pituitary gland

Review

Show what you learned in this unit. Answer these questions. The page where you can find the answer is listed after each question.

1. What system controls everything your body does? (page 127)
   N ________ system
2. What organs gather information about the world around you? (page 128)
   S ________ o ________
3. What organ in your body controls every action that your body does? (page 129)
   B ________
4. What system is made up of glands scattered throughout your body? (page 130)
   E ________ system
5. What chemicals make your cells do certain things? (page 130)
   H ________

Check These Out

1. For your Science Notebook
   a. *Science Work:*
      - Get a prepared slide of nerve cells. Look at it in a microscope. Draw what you see.
      - Get a cow's eye from a meat shop. Carefully cut it in half. Find these parts: the *retina, lens*, and *cornea*. Draw the eye and label those parts. If you need help, use a picture from a book, such as an encyclopedia.

   b. *Keep Fit:* Find out what *stress* is. Find out different ways a person can handle stress.

   c. *In the News:*
      - Drugs are chemicals that can help or harm your body. Find an article that tells what drugs can do to the body.
      - The brain has a left side and a right side. Find an article that talks about the sides of the brain.
2. Make a poster of the brain. Label the *cerebellum, cerebrum,* and *medulla.*
3. Sometimes endocrine glands don't make enough hormones. Pick an endocrine gland. Find out what happens when it doesn't make enough hormones.
4. Here are more things you may want to find out:
   - What are other endocrine glands? What does each gland control?
   - How does each sense organ work?
   - What is the nervous system like in the earthworm, dog, and whale?

Unit 6

Creating Life

Did you know this? Everyone begins life as *one cell*. From that one cell come the billions of cells that make up a person's body.

Males and females have organ systems that can start a new person—and a new life. If we didn't have those systems, the human race would die out. There would be no people in the world.

- What organ system starts life?
- When is a new life started?
- How does one cell turn into billions of cells?

You'll learn the answers in this unit.

Before You Start

You'll be using the science words below. Find out what they mean. Look them up in the Glossary. On a separate piece of paper, write what the words mean.

1. **fertilize**
2. **fetus**
3. **reproduce**

Starting from One Cell

All living things have a way to *reproduce* themselves. What do you think would happen if living things couldn't reproduce?

The human body has an organ system to reproduce humans. We call that system the **reproductive system**.

A baby is born with all the parts of his or her reproductive system. But that system doesn't start working until a person is around 12 years old. At that time, another organ system starts sending hormones to the reproductive system. What system sends the hormones?

Right! The endocrine system sends hormones to the reproductive system.

The hormones tell the reproductive system to start producing *sex cells*. In the female, those sex cells are called *eggs*. In the male, they are called *sperm*.

Those two kinds of sex cells cannot create a new life by themselves. To do that, they must join together and become one cell. When that happens, we say the sperm *fertilizes* the egg.

The picture shows an egg cell that has been fertilized by a sperm cell. How does that one cell become many cells?

The one cell divides and becomes two cells. The two cells divide and become four cells. The four cells divide, and so on.

As the cells divide, they begin to form tissues, organs, and the different parts of the body. Slowly, the cells build up into a baby. What is a baby inside its mother's body called?

Right! A baby inside its mother's body is called a **fetus.**

Many sperm cells come from the male's reproductive system. But only one sperm cell will fertilize the egg cell.

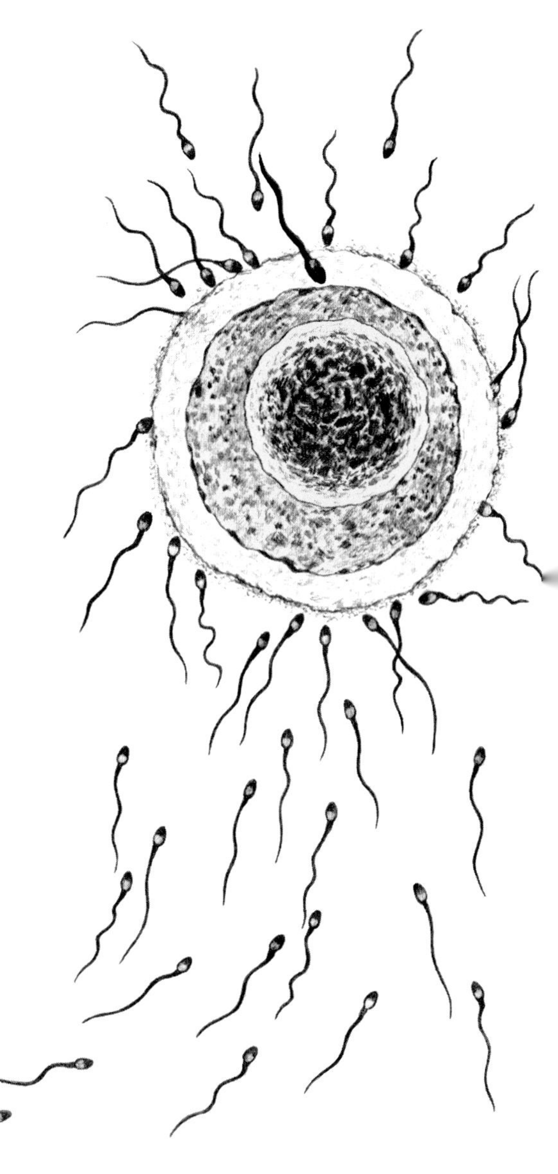

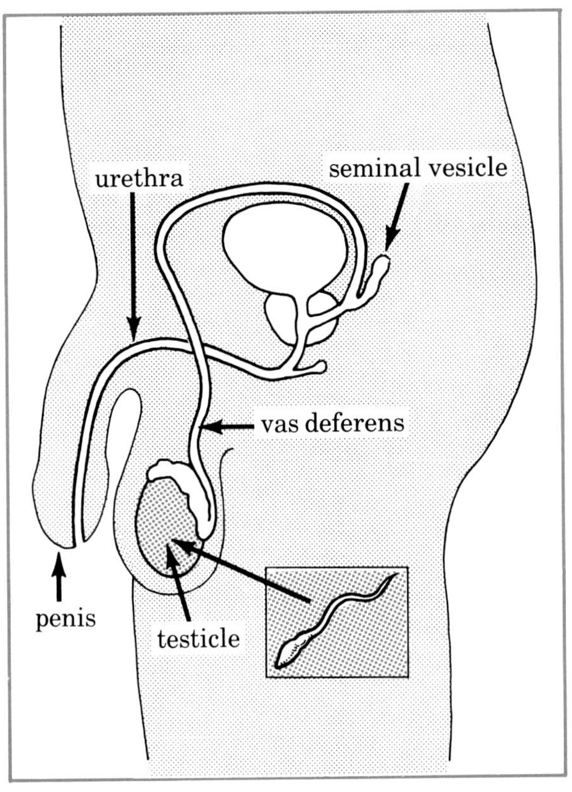

Male

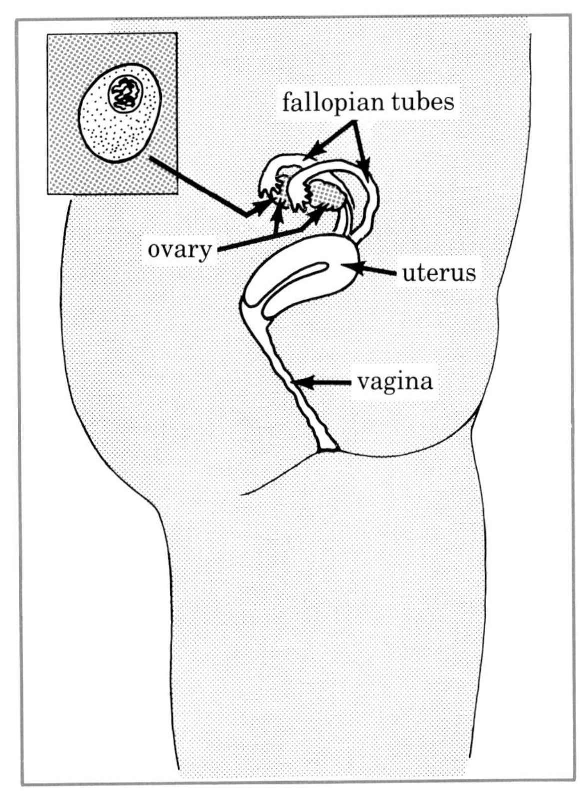

Female

The Reproductive Organs

The reproductive organs in a male and female are different. That's because they must do different things in order to produce a new human.

Look at the diagram of the male reproductive organs. What organ do you think makes the sperm cells?

The *testicle* makes the sperm cells. (The male body has two testicles.) Those sperm cells move through tubes called *vas deferens*. They go to the *seminal vesicle*. The sperm cells are stored there until they are ready to be released. Then they are pushed out of the body through the *urethra*. That is a tube in the *penis*.

Now look at the diagram of the female reproductive organs. What organ do you think produces egg cells?

The *ovary* produces egg cells. (The female body has two ovaries.)

The egg cell moves through a *fallopian tube*. If the egg cell joins a sperm cell along the way, it becomes fertilized. Then it attaches itself to the wall of the *uterus*. It slowly develops into a *fetus*.

When the fetus is ready to be born, we call it a *baby*. The baby leaves the uterus through the *vagina*.

A cell takes several months to develop into a baby that's ready to be born. How many months do you think it takes?

Nine Months

What day were you born? That day was your first day of life in the outside world! Before that day, you were developing inside your mother's uterus. You probably lived there for nine months. That's usually how long a human fetus lives inside its mother.

You learned that a person begins life as *one cell*. The cell multiplies and attaches itself to the wall of the uterus. At that stage, we call it an **embryo**. An embryo doesn't look like a human baby. What do you think it looks like?

The pictures on this page show how an embryo develops into a fetus. Look at picture **A**. It shows an embryo. The embryo is one month old. The heart has formed and is beating. (At this stage, the heart looks like a tube.) The brain and skeleton are just forming.

In picture **B**, the embryo is twice as big as it was in picture **A**. It is three months old. It now looks like a human baby. Now we call it a fetus. It has nerves, lungs, a stomach, intestines, a bladder, and muscles. But only the muscles are working. They let the fetus move in the uterus.

In picture **C**, the fetus is six months old. Its respiratory, circulatory, and digestive systems are almost completely formed. It moves around a lot. It kicks, turns, and swings its arms.

Picture **D** shows the fetus when it's nine months old. What do you think: Is the fetus ready to be born? Why?

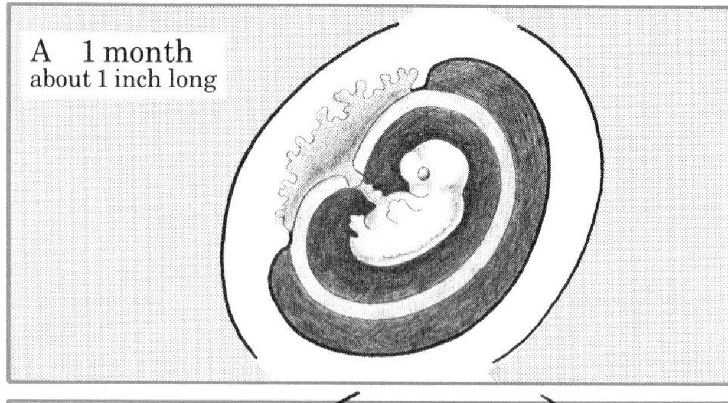

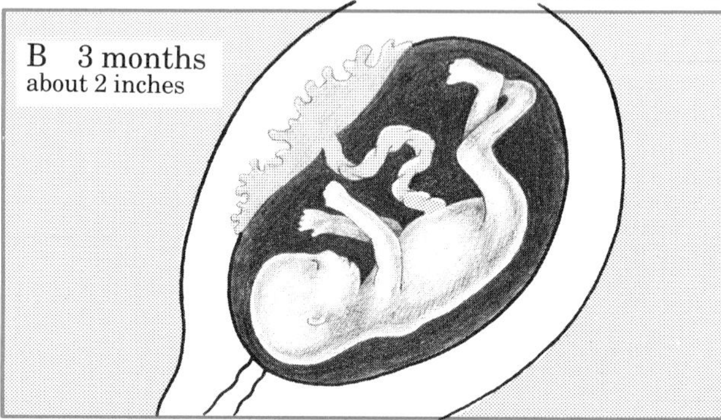

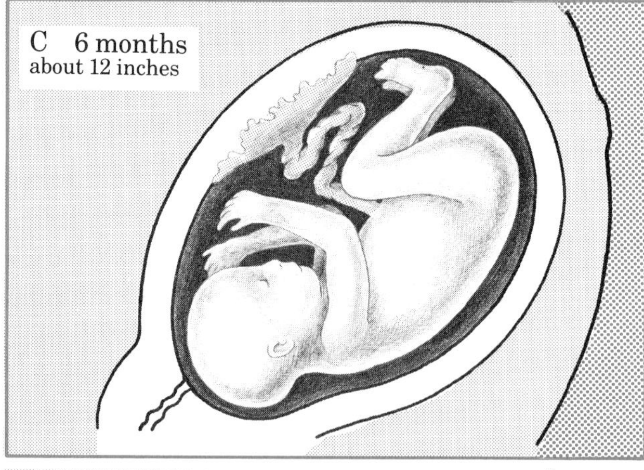

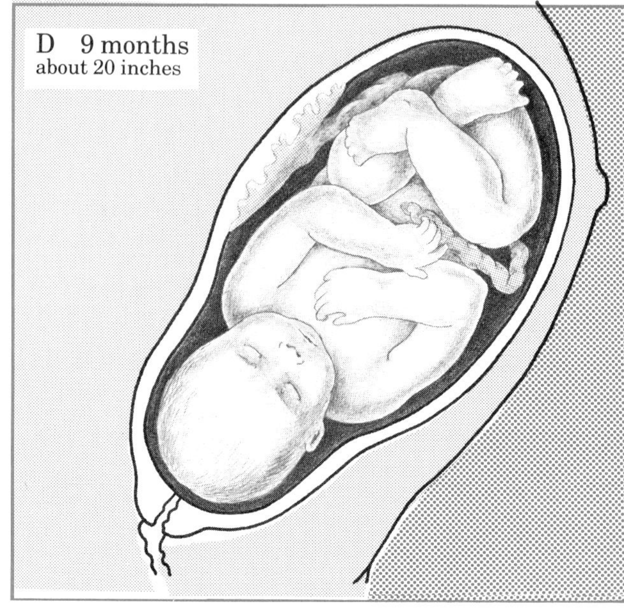

Review

Show what you learned in this unit. Use the words in the list below to answer the questions.

reproduce	fetus	egg
fertilize	sperm	uterus

1. What word means "to create a new living thing"?
2. What is an unborn baby called?
3. What word means "to join a sperm cell with an egg cell"?
4. What is the name of the place where a fetus grows?
5. What do you call a male sex cell?
6. What do you call a female sex cell?

Check These Out

1. For your Science Notebook:
   a. *Science Work:* Get a chicken egg. Carefully break open its shell. Look at the egg and its shell. Find these parts: the membrane, the albumen (white), and the part that holds the nutrients (yolk). Draw the egg and label its parts.
   b. *Keep Fit:*
      - A pregnant woman should eat certain foods to have a healthy baby. Find out what foods she should eat.
      - Doctors tell pregnant women not to smoke, drink alcohol, or take most kinds of drugs and medicines. Find out why.
   c. *In the News:* Sometimes women give birth to twins or triplets. These are called multiple births. Find an article about a multiple birth.
2. Make a "Baby Board." Have each student in class bring in a picture of himself or herself as a baby. Then put up the pictures on a bulletin board.
3. As a class, talk about the responsibilities a woman and a man might have as parents.
4. Here are more things you may want to find out:
   - What is the amnion? The placenta? The umbilical cord?
   - What is heredity?
   - How does a baby get Down's syndrome?

Unit 7

Taking Care

Eat Right!

You've learned about several of your organ systems. In order for your systems to work right, you must take care of them. What can you do to take good care of your systems?

One thing you can do is this: Eat foods that give your cells the nutrients, vitamins, and minerals they need. Here are some of them:

- Protein is a nutrient. It lets your cells grow and repair themselves. Protein is in foods such as hamburger, cheese, beans, and tofu. What are some other foods that can give you protein?
- Carbohydrate is another nutrient. It gives your cells fuel. Foods such as rice, tomatoes, and apples have carbohydrate. What are other foods that have carbohydrate?
- Your body needs many different vitamins. Vitamin C is just one of them. Fresh fruit and vegetables are rich in vitamin C. What other foods can you eat to get vitamin C?
- Calcium is a mineral everyone needs. It keeps the skeleton strong. Milk and ice cream are two foods rich in calcium. What are some other foods that give you calcium?

These are only a few of the nutrients, vitamins, and minerals your body needs. There are many more. If your body doesn't get enough of any of them, you get sick. What are the other things your body needs? What foods give you those things? Find out!

Watch Out!

Sometimes people eat, drink, and breathe things that can damage their bodies. They smoke too many cigarettes. They drink too much alcohol. Or they abuse other drugs such as marijuana, cocaine, and heroin.

Cigarettes, alcohol, and other drugs contain strong chemicals. Those chemicals can harm the body in many ways. Here are some ways:

Cigarettes contain a chemical called *nicotine.* (Nicotine is often put in sprays that are used to kill bugs.) Find out how nicotine can damage the respiratory system.

Drinking too much alcohol can damage the liver. Find the answers to these questions:

What is the liver's job in the human body?

What happens when alcohol damages the liver?

Some other drugs are very harmful. Many scientists now think that some drugs can destroy certain nerve cells in the brain. They think that marijuana, LSD, and PCP (angel dust) can destroy the nerve cells that help a person think. How could that be bad for a person?

Show What You Learned

What's the Answer?

Below is a list of the organ systems you have learned about. Find the right organ system for each set of words that tells what that system does.

muscular	nervous
excretory	endocrine
lymphatic	reproductive
respiratory	digestive
skeletal	circulatory

1. Moves oxygen, nutrients, and waste products through the body
2. Controls how the body acts
3. Moves oxygen and carbon dioxide in and out of the body
4. Breaks down food into nutrients
5. Can create another life
6. Moves body parts
7. Traps and kills germs and bacteria
8. Controls the body with hormones
9. Gives your body support
10. Removes liquid waste

What's the Word?

Give the correct words for the meanings.

1. Parts of food the cells use
   N ________
2. A chemical that tells the body to act a certain way
   H ________
3. The smallest living thing
   C ________
4. The way the bones are put together to support the body
   S ________
5. An unborn baby
   F ________
6. The gas the body needs to live
   O ________

Congratulations!

You've learned a lot about the human body. You've learned

- What ten organ systems are
- What each organ system does
- What things keep your body alive and healthy
- And many other important facts about the organ systems

THE FIVE SENSES

What kinds of senses do living things have? How do we use our senses to stay alive? What are the *five senses*? How do they work? In this section, you'll learn many facts about the five senses. And you'll learn how they play an important part in our lives.

Contents

Introduction

Suppose someone gives you a bag of chips. You take a chip and eat it. What will you do if the chip tastes good? What will you do if the chip tastes bad?

If the chip tastes good, you'll probably eat more. If it tastes bad, you'll probably stop eating. You may even throw the bag of chips away.

Whatever you do depends on what your sense of taste tells you.

Taste is one of the senses that our bodies use to find out about the world around us. We call those senses *the five senses*.

The five senses gather facts for our bodies. They tell us what's happening around us. They help us stay alive and well.

Imagine what it would be like if we didn't have the five senses. We wouldn't know what things sound and feel like. We wouldn't know when we are in danger. Or when something happens that can make us happy.

This section will help you learn about the five senses. You'll learn what they are and how they work. You'll do interesting science activities and experiments.

And you'll find out what the senses in *your* body can do.

Unit 1

The Five Senses

What do you know about potato chips? You may know what they look, smell, and feel like. And you may know how they sound and taste when you bite into them.

If you know any of those things, it's because you've learned about potato chips. You've learned about them through your senses.

We call the senses that we learn through *the five senses*. Everything that you know, you've learned through some or all of those senses. You're using some of them right now to learn about the five senses!

- What are the five senses?
- How do the five senses work?

You'll find out in this unit.

Before You Start

You'll be using the science words below. Find out what they mean. Look them up in the Glossary that's at the back of this book. On a separate piece of paper, write what the words mean.

1. **connected**
2. **information**
3. **sensitive**

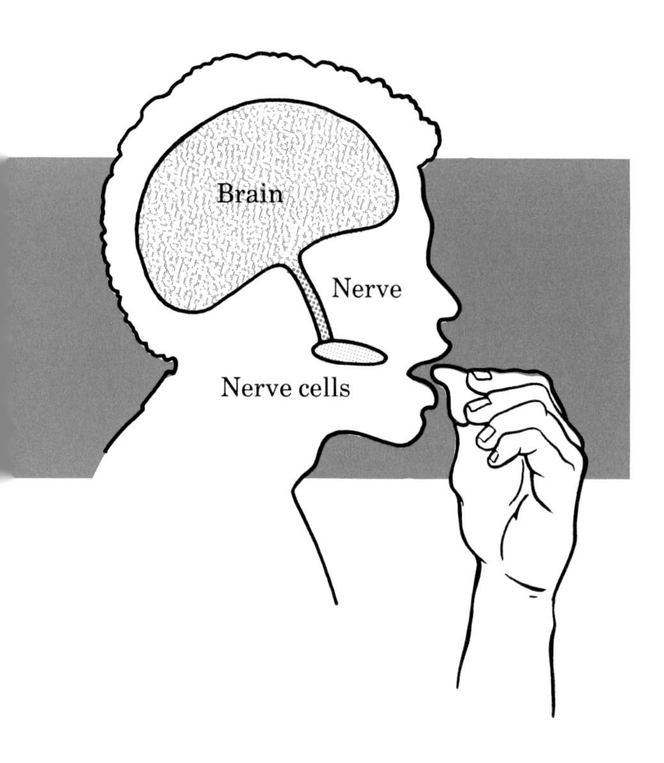

Getting the Facts

There's a part of your body that controls everything you do. For example, it makes you eat something. It makes you like or not like what you are eating. It makes you eat more or stop eating.

What is that part of your body?

Right! That part of your body is your **brain**.

You act in many different ways. How you act depends on information your brain gets about everything around you. How does your brain get that information?

Right! Your brain gets information through your senses.

Here's how your senses work:

Your body has parts in it called **nerves**. Nerves are connected to the brain. At the ends of the nerves are parts called **nerve cells**. Nerve cells are sensitive to certain things. For example, some of your nerve cells are sensitive to how things taste.

Nerve cells get information about things such as taste. They pass that information along the nerves to the brain.

Our bodies have many senses. But the senses people usually talk about are the five senses. What do you think the five senses are?

Our Sense Organs

The five senses are *hearing, sight, smell, taste,* and *touch.*

You learned that nerve cells send messages about each of the five senses. Those nerve cells are found in certain parts of the body. Those parts are called **sense organs**.

Each sense organ sends only one kind of message. For example, one sense organ sends only messages about what things look like. Another sends only messages about how things taste.

What is the sense organ for each of the five senses? Look at the pictures on the next page. Then answer the questions.

1. What is the sense organ for hearing?
2. What is the sense organ for sight?
3. What is the sense organ for smell?
4. What is the sense organ for taste?
5. What is the sense organ for touch?

Use Your Senses!

What can your senses tell you about a soft drink? Pour a soft drink into a glass. *Listen* to the sounds the drink makes. Then *look at, smell, taste,* and *touch* the drink.

Then answer the questions about what you hear, see, smell, taste, and feel.

The five senses

1. **Hearing**
   - What sound does the drink make?
   - How loud is it?
   - Does it sound like anything else you can think of?

2. **Sight**
   - What does the drink look like?
   - What color is it?
   - Is it dark or can you see through it?

3. **Smell**
   - What does the drink smell like?
   - Is the smell strong or weak?
   - Does it smell like anything else you can think of?

4. **Taste**
   - What does the drink taste like?
   - Does it taste sweet?
   - Do you like the taste?

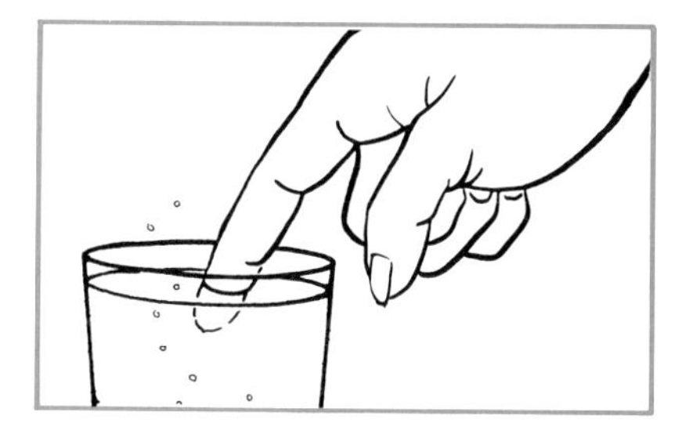

5. **Touch**
   - What does the drink feel like?
   - Is it sticky?
   - Is it cold or warm?

Review

Show what you learned in this unit. Use the words in the list below to finish the sentences. One word will be left over.

brain	five	information
nerves	sense organ	smells
sounds	tastes	touch

1. The ________ controls everything we do.
2. The brain gets ________ from our senses.
3. ________ carry messages to the brain.
4. The skin is sensitive to ________.
5. Nerve cells in the nose send messages about ________.
6. Nerve cells on the tongue send messages about ________.
7. Nerve cells in the ear send messages about ________.
8. The eye is the ________ ________ for sight.

Check These Out

1. Make a Science Notebook for this section. Use it to keep a record of what you learn about the five senses. Put your list of glossary words and their meanings in the notebook. Also keep a record of the experiments you do. You can put anything else you learn about the five senses in your notebook too.
2. You use your five senses even when you're asleep. How do you use each sense when you're sleeping? Find out.
3. Some people do not have all of the five senses. For example, a person who is blind does not have the sense of sight. But that person learns to use some other sense or senses to make up for the missing sense. So a blind person may have a very sharp sense of hearing, or a very sensitive sense of touch. Find out how a deaf person can make up for the missing sense of hearing.
4. Alcohol and other drugs can cause the senses not to work right. Find out what happens to the senses when a person uses alcohol or other drugs.
5. As you work your way through this section, you may want to find out more about the five senses. You can find out more by looking in an encyclopedia or by getting books about the five senses from a library. You can also talk to an expert, such as a doctor or a science teacher.

   Here are some things you may want to find out:
   - What is the nervous system? How does it work?
   - What are the external senses? What are the internal senses?
   - At what age does a person get each of the five senses?
   - What is the "sixth sense"? How is it different from the five senses?

Unit 2

The Sense of Sight

Your eye is like a very good camera. It can take pictures that are still or moving, in color or in black and white. It can take close-up pictures of a tiny thing. It can take wide pictures of huge things.

But look at the pictures a real camera takes. And think of the pictures your eye sees. The pictures your eye sees are much better than those a camera takes! No camera has yet been made that's better than the eye.

- How does the eye work?
- How does the sense of sight work?

You'll find out in this unit.

Before You Start

You'll be using the science words below. Find out what they mean. Look them up in the Glossary. On a separate piece of paper, write what the words mean.

1. **enters**
2. **focus**
3. **image**

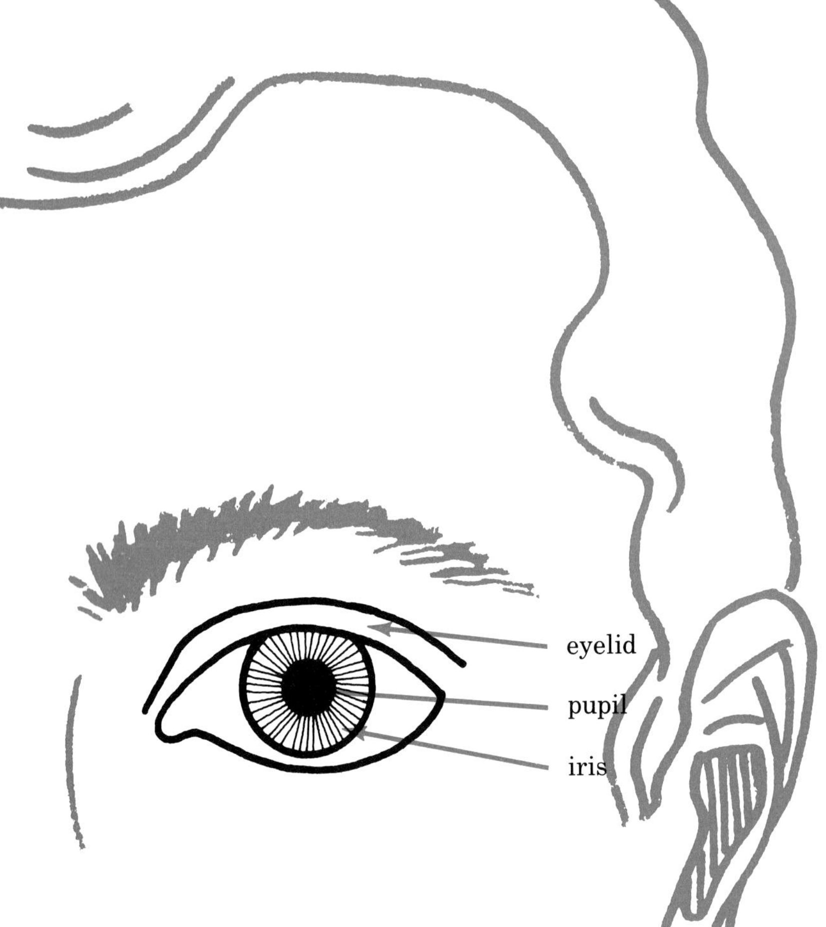

The Outside of the Eye

As you read this, your eyes are taking pictures of words. Close your eyes a little, and keep on reading. What happens?

Right! You can't see the words very well. When you close your eyes, you keep some light out of your eyes.

In order for you to see, light must enter your eyes. But just the right amount of light must enter. So the outside of each eye has a part that lets the right amount of light in.

Get a mirror and look at your eyes. Look at the **iris** in each eye. The iris is the big, colored circle in the eye. What color is your iris?

Now look at the smaller, dark circle in the middle of each iris. That smaller circle is called a **pupil**. The pupil is really a hole in the iris. Light enters the eye through that hole.

The iris is a muscle. It can make the pupil bigger or smaller. It does that to control the amount of light that enters the eye.

One other part controls the amount of light that enters the eye. It does that by covering the eye. What is that part?

Yes! That part is the **eyelid**. The eyelid can be lowered to keep out light. It can be raised high to let in more light.

Look at the picture of the eye on this page. It shows the three parts you just learned about. Find the iris, the pupil, and the eyelid.

Experiment 1

What happens to the eye in dim light and bright light?

Materials (What you need)

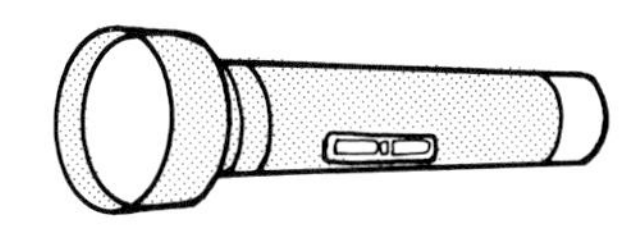

One small mirror

One flashlight

Procedure (What you do)

1. Sit in a place where the light is dim. Look into the mirror at your eyes. Are the pupils big or small?

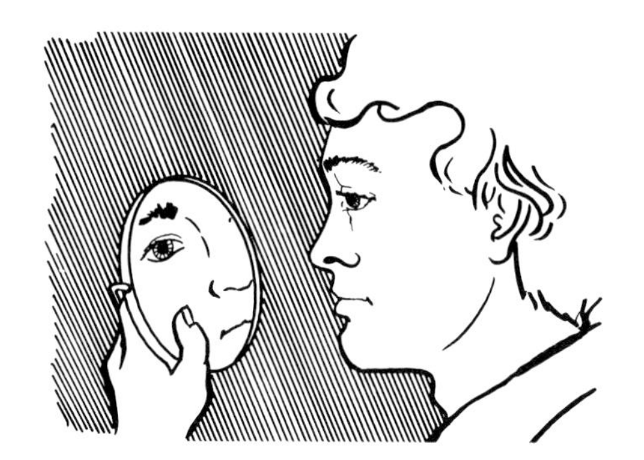

2. Shine the flashlight on your face. The light near your eyes is now bright. Are the pupils big or small?

Observations (What you see)

1. When the light is dim, are the pupils big or small?
2. When the light is bright, are the pupils big or small?

Conclusions (What you learn)

Finish these sentences. Use the words *bright, dim, big,* and *small.*

1. When light is ________, not enough light can enter the eye. The pupils become ________.
2. When light is ________, too much light can enter the eye. The pupils become ________.

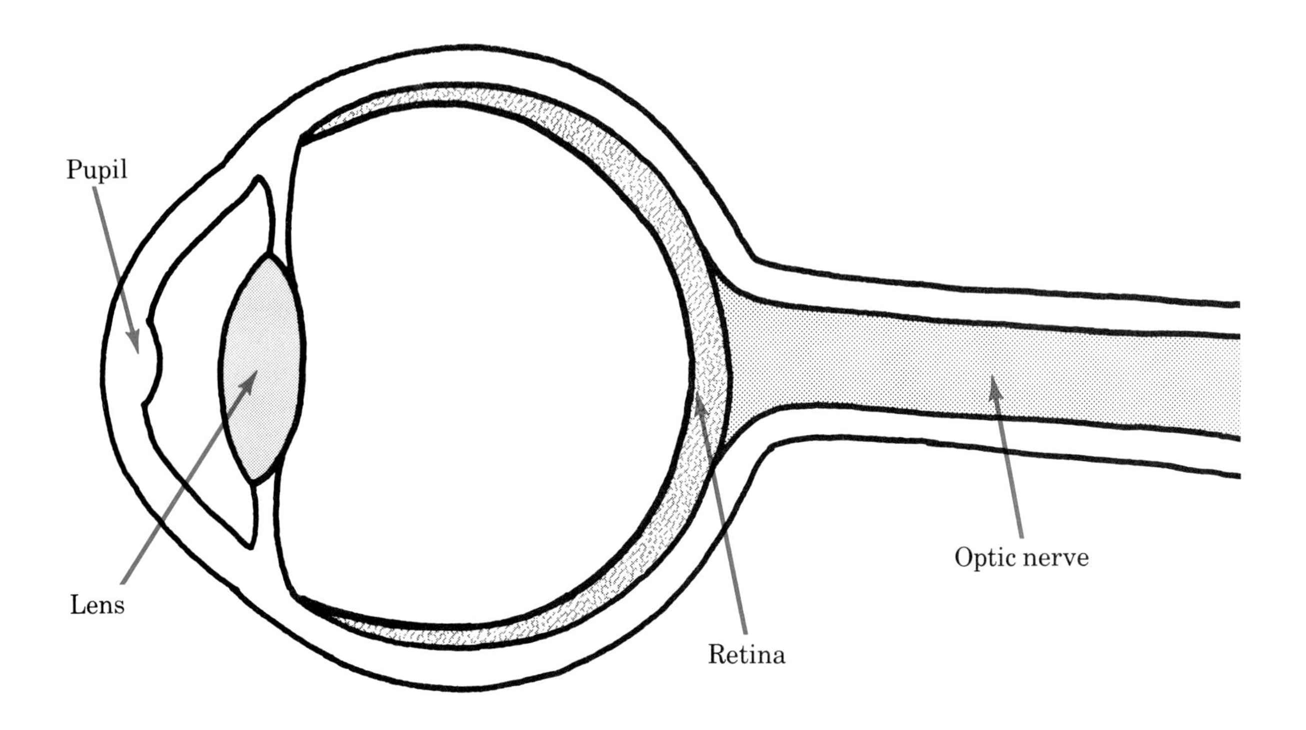

Inside the Eye

You see objects in the world—trees, people, buildings. They seem "out there." But what you're actually seeing is in your head!

What happens is this: Light rays bounce off objects in the world. The light rays carry the color, shape, and texture of the objects that they bounce from.

Let's say you're looking at a chair. Light rays from the chair travel to your eyes. They enter through the pupils of your eyes. Next, they travel through a part of each eye called the **lens**. The lens gathers together the light rays and sends them to the back of the eyeball. They land on the **retina**. There, the light rays form an *image* of the chair. An image is like the picture on a movie screen—it is made by light rays.

Just like a movie-screen image, the image on the retina must be in focus. This is the job of the lens. If the light rays come from a distant chair, the lens flattens out. If the light rays come from a chair close by, the lens thickens. In this way, the image of the chair forms on the retina.

The retina has thousands of different nerve cells. These cells are sensitive to light. Each cell senses a part of the image. It sends messages about this part to your brain. These messages are sent along the **optic nerve**.

Your brain gets the messages from the retina cells. It puts them together. It makes a mental picture of the chair. That mental picture has the color, shape, and texture of the actual chair.

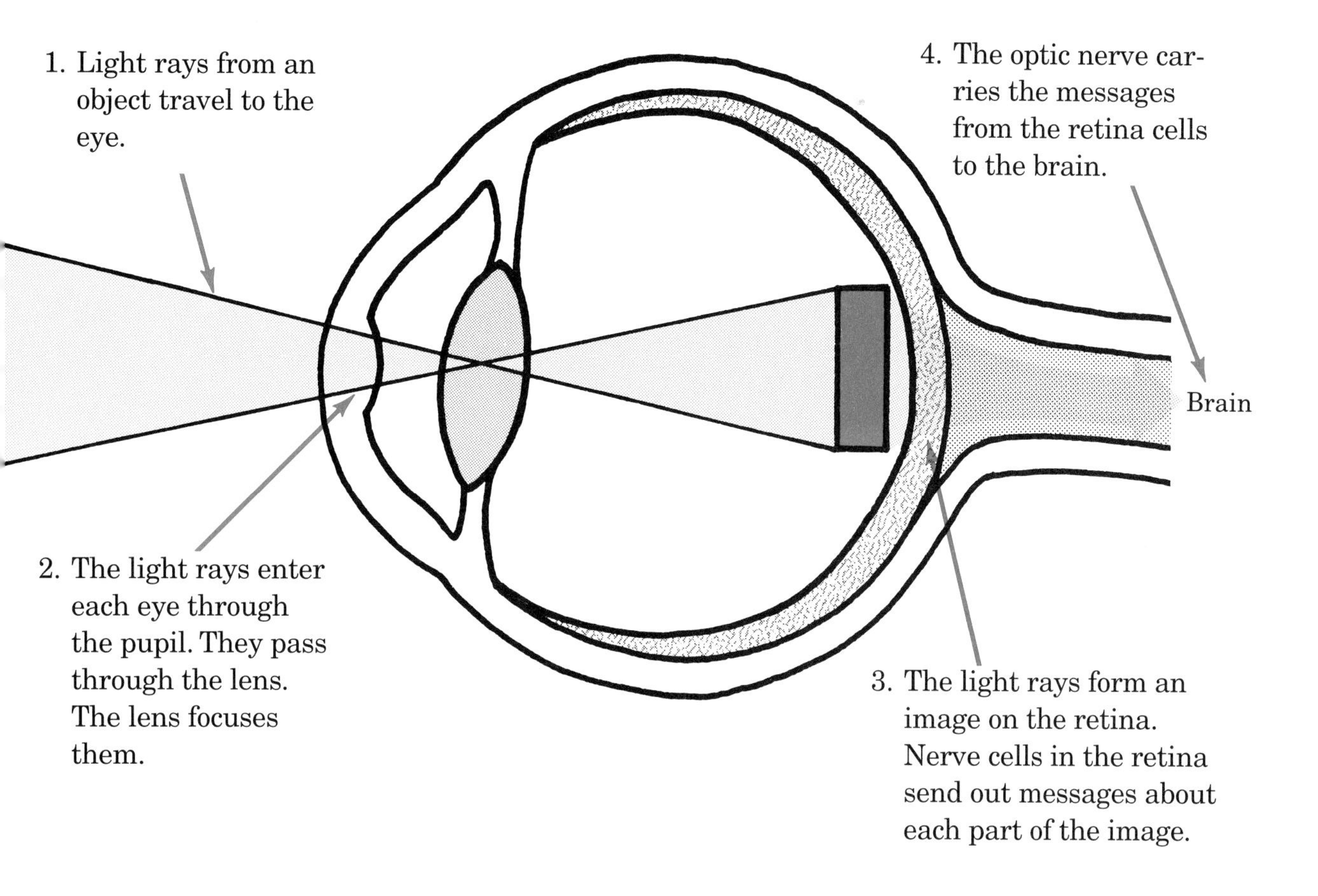

1. Light rays from an object travel to the eye.
4. The optic nerve carries the messages from the retina cells to the brain.
Brain
2. The light rays enter each eye through the pupil. They pass through the lens. The lens focuses them.
3. The light rays form an image on the retina. Nerve cells in the retina send out messages about each part of the image.

How the Sense of Sight Works

The diagram on this page shows how you see. Read the sentences in the diagram. They describe what happens inside the eye when you look at something. Then answer the questions below.

1. When you look at an object, what travels from the object to your eyes?
2. Light rays enter each eye through what? What focuses the light rays?
3. Where do the light rays form an image in the eyeball? Nerve cells in that part send messages about the image. What nerve do those messages travel along?
4. The optic nerve sends the messages to what part of the body?

The brain puts the messages together, and you see.

Experiment 2

What kind of image is formed on the retina?

Materials

One cardboard box

One magnifying glass

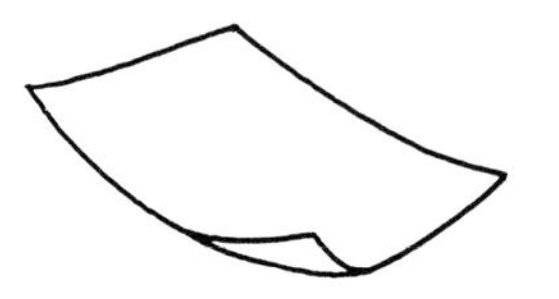

One piece of white paper

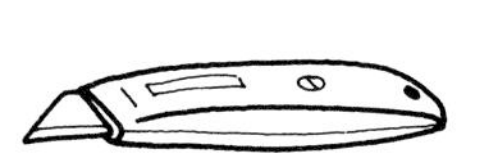

One knife (or a pair of scissors)

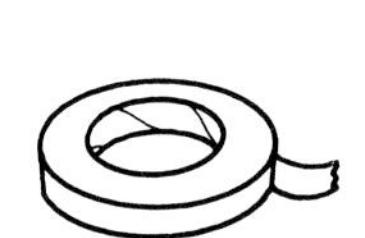

Tape

One lamp

One large picture on white paper

Procedure

1. Cut a hole in the bottom of the box. Make the hole a little smaller than the magnifying glass.

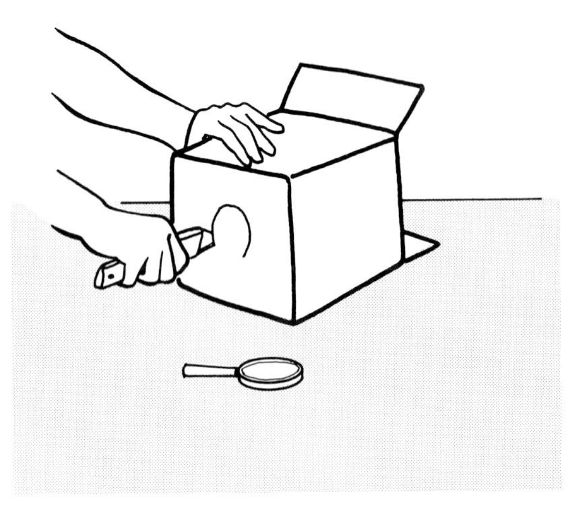

2. Tape the magnifying glass over the hole.

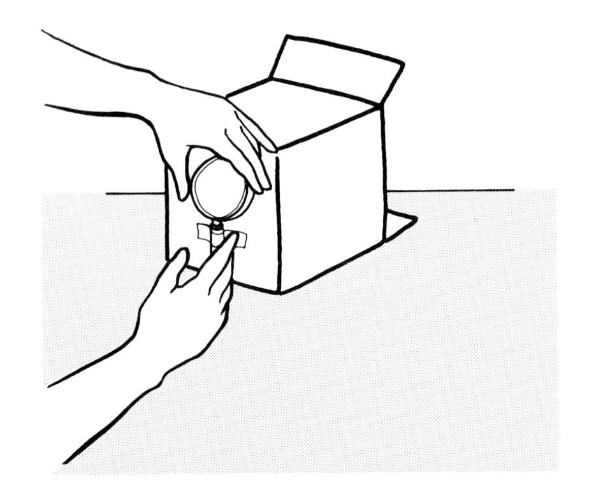

3. Hold the picture so it faces the open end of the box. Tape the picture over that end. Put the lamp next to the back of the picture. Turn the lamp on.

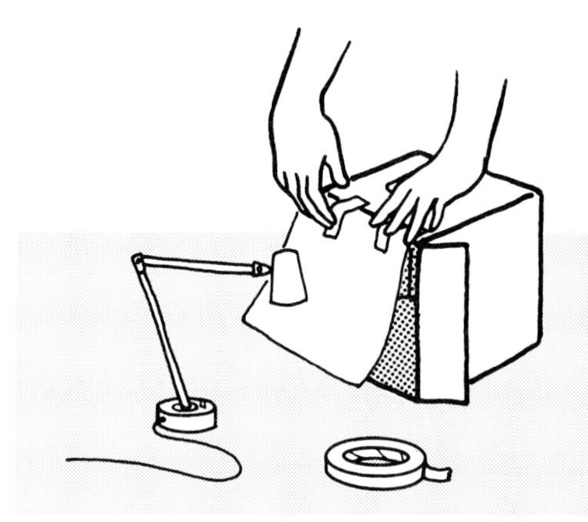

4. Turn off all the lights in the room. Hold the white paper in front of the magnifying glass. Slowly move the paper away from the box. What do you see on the paper?

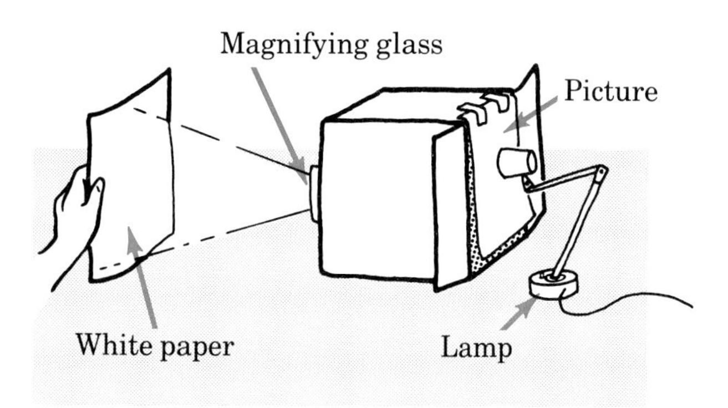

Observations

What do you see on the white paper?

You should see an upside-down picture on the paper. That picture is the same as the one you taped on the box.

Conclusions

1. The hole in the box, the magnifying glass, and the paper are like parts in your eye. The hole is like the pupil. The magnifying glass is like the lens. What is the paper like?

   Right! The paper is like the retina.
2. What kind of image is formed on the retina?
3. Why don't we see things upside down?

Experiment 3

How do two eyes help you see better?

A camera takes pictures. Your eyes take pictures too.

But a camera takes one picture at a time. You have two eyes. So your eyes take two pictures at a time. Your brain puts those two pictures together.

You see better with two eyes than if you had only one eye. How? Do this experiment and find out.

Materials

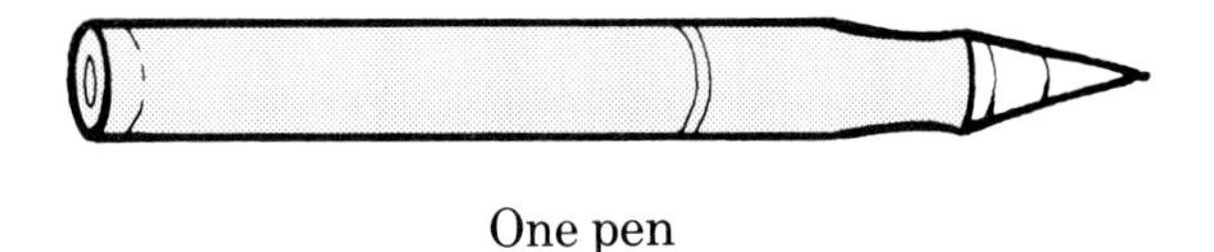

One pen

The cap for the pen

Procedure

1. Hold the pen in one hand. Hold it straight out in front of you. Hold the cap in your other hand.

2. Close one eye. Keep your arms straight. Put the cap on the pen. Is it easy to do?

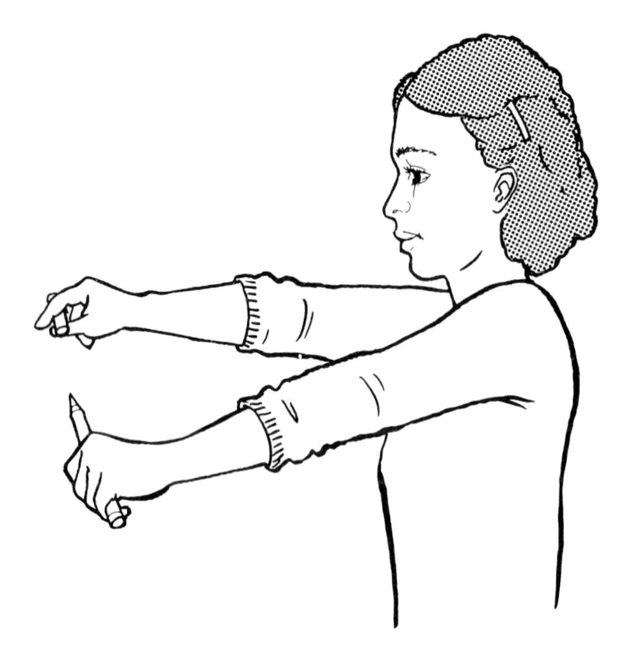

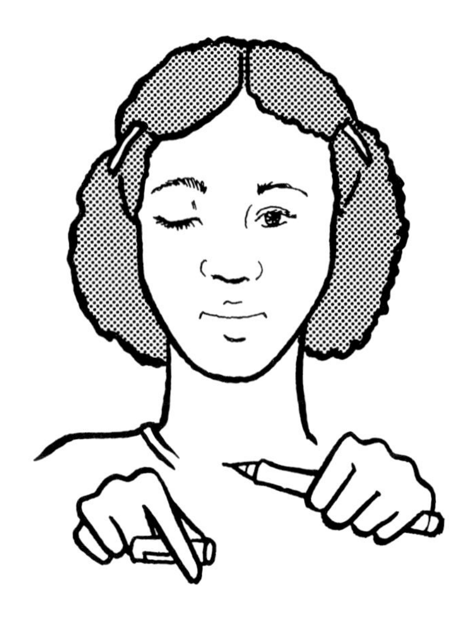

3. Now take the cap off the pen. Keep both eyes open. Put the cap on the pen again. Is it easier to do this time?

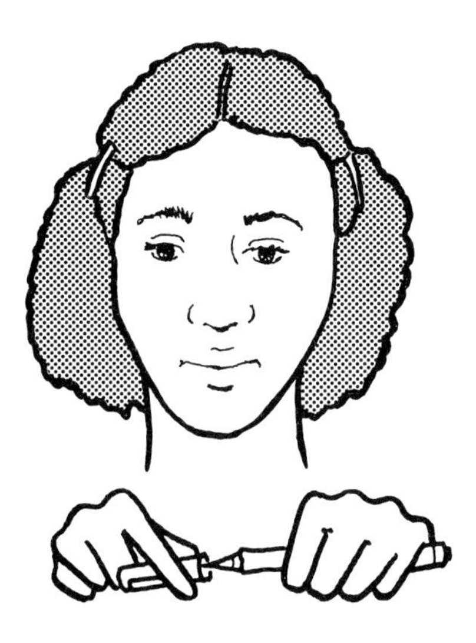

Observations

When is it easier to put the cap on—with one eye closed, or with both eyes open?

Conclusions

1. Why is it harder to put the cap on with one eye closed?

   Right! It's harder because you don't know exactly how far away the pen is.
2. Why is it easier to put the cap on with both eyes open?
3. How do two eyes help you see better?

   Right! Two eyes help you know how far away something is.
4. Two eyes help you do things, such as step down from a high place to a lower place. What's another example of things two eyes help you do?

Review

Use the words you learned in this unit to answer the questions. The words you'll need are listed below.

iris	light rays	pupil
retina	focuses	brain
lens		

1. On what part of the eye do the light rays form an image?
2. The lens gathers together the light rays on the back of the eye. What's another word for "gathers together"?
3. What part of the eye becomes larger in dim light and smaller in bright light?
4. In what form does the picture come to the eye?
5. What part of the eye is a muscle that lets in just the right amount of light?
6. What part helps us see things near and far away?
7. The image inside the eye is upside down. What makes the image look right-side up?

Check These Out

1. Find out how a camera takes a picture. Then make a poster that shows a camera and an eye. On one half of the poster, show how a camera takes a picture. On the other half, show how an eye takes a picture.
2. How well can you see? When was the last time you had your eyes checked? Invite your school nurse or an eye doctor to your class. Have that person bring an *eye chart*. Use the eye chart to check how well you can see.
3. To have good sight, you must take care of your eyes. Find out how to take care of your eyes. Then make a list of what to do and what not to do. Put the list up in class.
4. Here are more things you may want to find out:
   - You learned about three parts in the front of the eye. What are the other parts in the front of the eye? What do those parts do?
   - Many people are nearsighted or farsighted or have astigmatism. What do those things mean? What has to be done to the eye to fix each one?
   - How do we see color? What is color blindness?
   - What's your dominant eye?
   - Who is Stevie Wonder? What does he do?

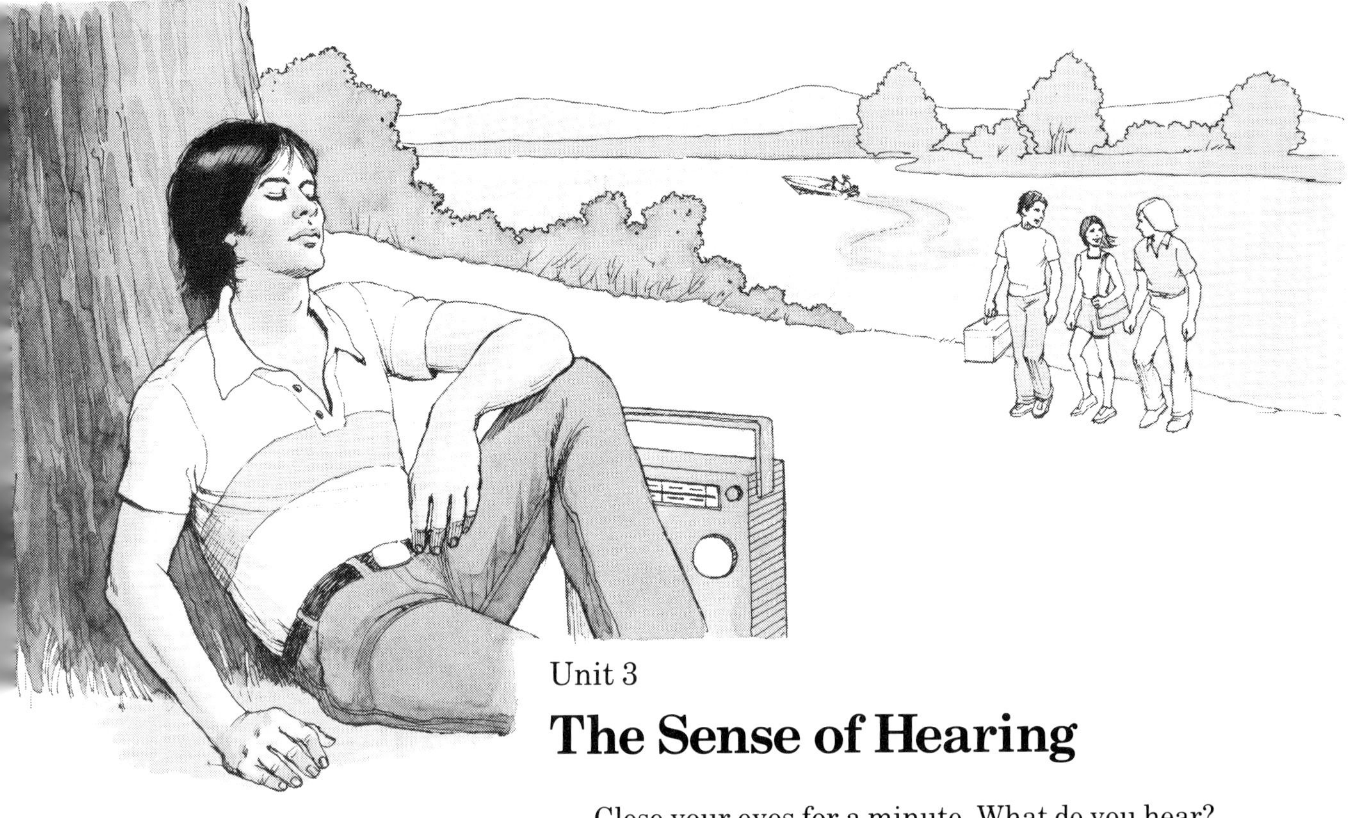

Unit 3

The Sense of Hearing

Close your eyes for a minute. What do you hear? Do you hear people talking? Do you hear things moving?

Every second of the day, our ears hear things. They hear soft sounds as well as loud sounds.

Our sense of hearing helps us find out what's going on around us.

- How do we hear things?
- How does the sense of hearing work?

You'll find out in this unit.

Before You Start

You'll be using the science words below. Find out what they mean. Look them up in the Glossary. On a separate piece of paper, write what the words mean.

1. **auditory nerve**
2. **stimulate**
3. **vibration**

Why We Hear

In Unit 1 of this section, you learned what the sense organ for hearing is. What is that sense organ?

Right! The sense organ for hearing is the ear. Nerve cells deep inside the ear are sensitive to sounds that things make.

Here's how something makes a sound:

- The thing **vibrates**—it moves back and forth very quickly.
- Its vibrations cause vibrations in the air. The vibrations in the air are called **sound waves**.
- The sound waves travel to your ear.
- The sound waves enter your ear, and you hear the sound.

You can see how something vibrates and makes a sound. Do this:

Get a rubber band. Stretch it between two fingers. With your other hand, pull on the rubber band and let it go. You'll see it vibrate, and you'll hear the sound it makes.

Some things make music the way the rubber band makes sound. Name some of those things that make music.

Vibrating Things

You learned that when something vibrates, it moves back and forth quickly. Suppose that vibrating thing is touching something else. What do you think the second thing will do?

Yes! The second thing will also vibrate.

You can see for yourself how a vibrating thing makes another thing vibrate. You'll need these things:

- One radio
- One clear plastic glass with a little water in it

---

1 Turn on the radio. Turn it up loud. Put your hand on the radio. What do you feel?

Right! You feel the radio vibrate.

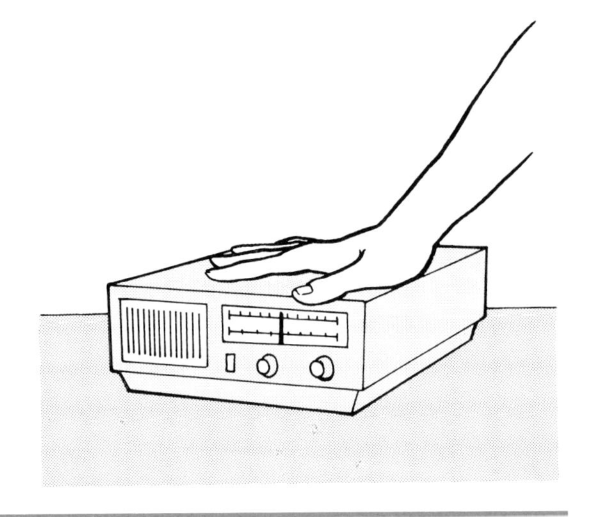

---

2 Now put the glass of water on top of the radio. Watch the water. What does it do?

Right! The water moves. It moves because the glass vibrates.

What makes the glass vibrate?

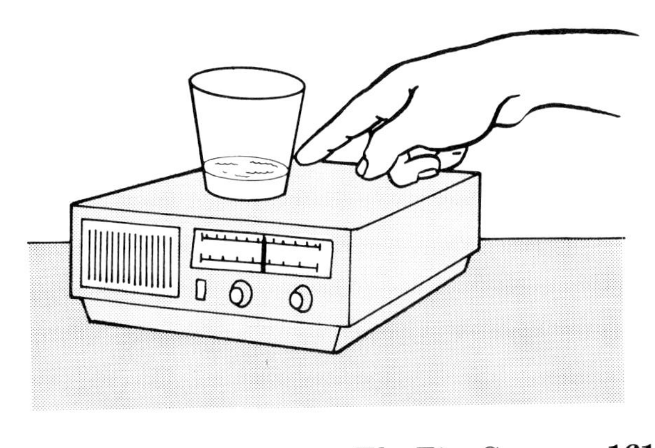

How the Sense of Hearing Works

Let's say you're listening to a band. The music makes vibrations in the air—sound waves. Those sound waves travel to your ear.

Look at the diagram on the next page. It shows what happens when sound waves reach the ear.

The sentences below tell what happens to the sound waves. But some of the words are missing. The missing words are the names of parts of the ear. You can find those names on the diagram. Find the names and use them to finish the sentences.

1. Sound waves enter through a hole in the ________ ________.
2. Sound waves hit the ________ in the middle ear. That part vibrates.
3. The vibrating eardrum makes three small ________ vibrate.
4. The vibrating bones make parts in the ________ ________ vibrate. Nerve cells are stimulated.
5. The nerve cells send a message along the ________ ________.

The message goes to the brain, and you hear a sound.

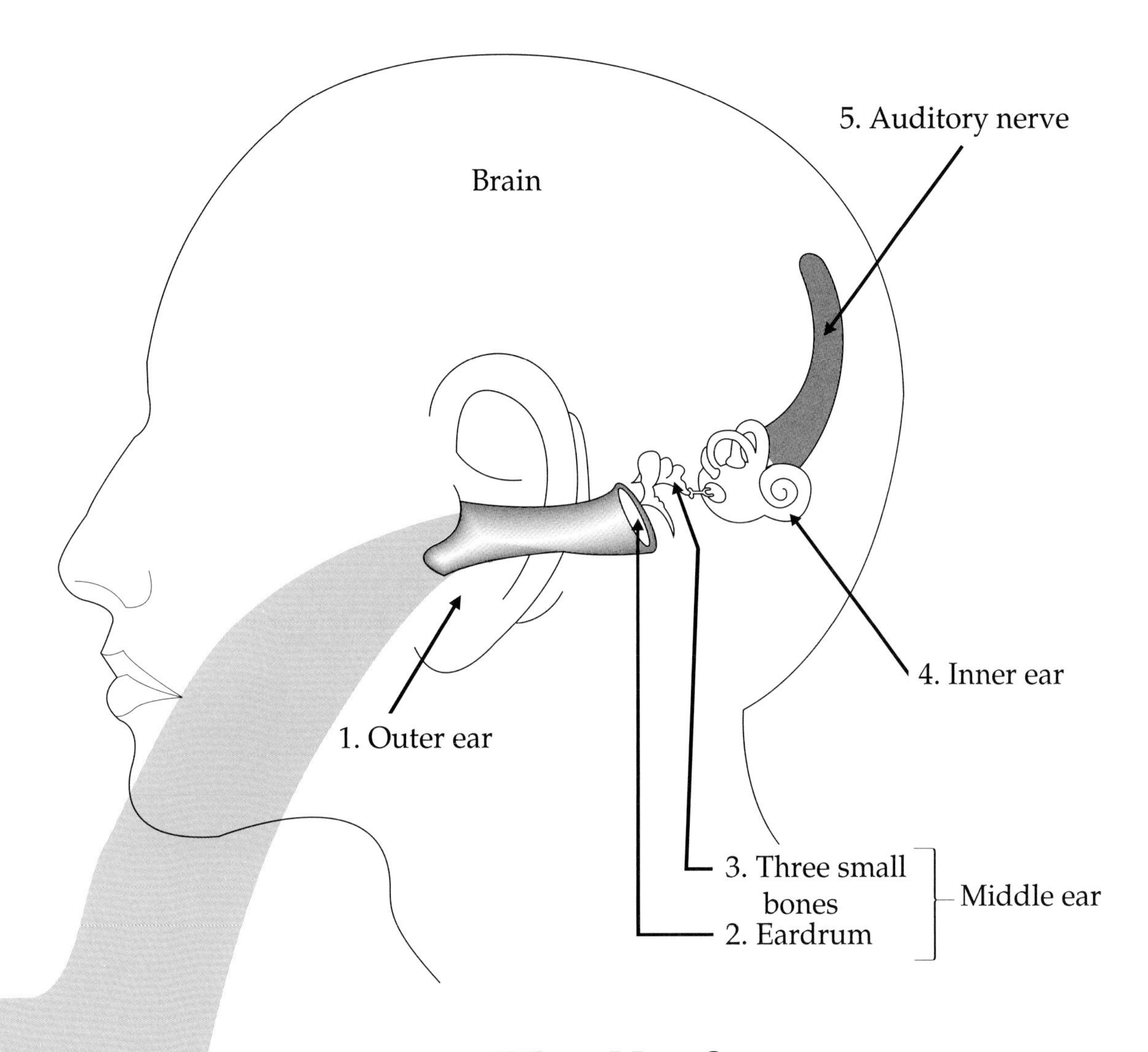

What Next?

When something makes a sound, sound waves enter the ear. Each part in the ear makes the next part vibrate. Look at the diagram. Then answer these questions.

1. What part of the ear vibrates first?
2. What part vibrates next?
3. Then what part vibrates?
4. The vibrations stimulate nerve cells. What do the nerve cells do?

The nerve cells send the message along the auditory nerve. That message goes to the brain. And a sound is heard.

What Do You Hear?

Our sense of hearing lets us hear many different sounds. We hear loud sounds and soft sounds. We hear sounds that are close and sounds that are far away. We hear sounds that people make and sounds that things make.

Find out what sounds you can hear right now. Sit next to an open door or window. Cover your eyes with your hands so that you can see only this page. Then listen carefully and answer these questions.

1. What is the loudest sound you hear?
2. What is the softest sound you hear?
3. What is the closest sound you hear?
4. What is the farthest sound you hear?
5. Do you hear someone talking? If you do, what is that person saying? Where is that person?
6. Do you hear cars and trucks passing by? If you do, are they going fast or slow?
7. Do you hear a machine being used? If you do, what is it? Where is it?
8. Do you hear any sounds that tell you about the weather? If you do, what are they?

What's Going On?

You use your sense of hearing to find out what's going on around you. Sometimes you use it to find out what another person is doing. Try it. You'll need a partner to do the things below. Don't look at your partner. Guess what your partner is doing from the sounds you hear.

1 Your partner will use a book to make a sound. (Examples: Drop the book on the floor. Turn the pages. Slam the book shut.) What is your partner doing with the book?

2 Your partner will choose three different things in the room. Your partner will use them to make sounds. What three things is your partner using?

3 Your partner will talk to you from somewhere in the room. In what part of the room is your partner?

4 Your partner will stand about four feet away from you. Then your partner will walk toward you or away from you. Which way is your partner walking?

Your sense of hearing helps you find out many things. It helps you find out what someone is doing. And it helps you find out what's making a sound. It also helps you find out where someone is. So when you want to know what's going on, you can use your sense of hearing.

Review

Show what you learned in this unit. Finish these sentences. Match the first part of the sentence on the left with the correct words on the right.

1. To make a sound, something vibrates
2. Sound waves travel from
3. When sound waves enter the ear,
4. Vibrations in the inner ear
5. Nerve cells in the inner ear send a
6. The message goes

a. they make parts in the ear vibrate.
b. the vibrating thing to the ear.
c. and causes sound waves.
d. message along the auditory nerve.
e. to the brain, and a sound is heard.
f. stimulate nerve cells.

Check These Out

1. A person who is born without a sense of hearing has a hard time learning to talk. Find out why.
2. Some deaf people use sign language to talk with other people. Find a book about sign language in a library. Learn some of the signs.
3. Some deaf people learn to read other people's lips. Try it yourself. Watch a TV show, but turn off the sound. See if you can understand what the people in the show are saying.
4. Try to find a show by the National Theater of the Deaf on TV. Or find a show where someone *signs* the words for deaf people who are watching. Tell the class about the shows you find.
5. Here are more things you may want to find out:
   - Who was Helen Keller? Who was Annie Sullivan? Find out about their lives.
   - How do hearing aids work?
   - What are some dos and don'ts for taking care of your ears?
   - How do your ears help you keep your balance?

Unit 4

The Sense of Smell

Suppose you're ironing a shirt in the kitchen. You leave the kitchen to answer the phone. While you're on the phone, you notice a bad smell coming from the kitchen. That smell tells you this: You forgot to take the iron off the shirt, and the shirt is starting to burn!

We learn a lot about our world through smells. Smells help us know what's happening. And they help us know about the things around us.

- Why do we smell things?
- How does our sense of smell work?

You'll find out in this unit.

Before You Start

You'll be using the science words below. Find out what they mean. Look them up in the Glossary. On a separate piece of paper, write what the words mean.

1. **molecules**
2. **nostrils**
3. **olfactory nerves**

Smells Tell About Things

Imagine this: It's dinner-time. You come home and smell something cooking. The smell tells you that someone is frying chicken. The smell also tells you that the chicken is almost ready to eat.

You're using your sense of smell to tell you what you're having for dinner and when you'll be eating.

Most foods and many other things have an **odor**—a smell. We have a sense organ that is sensitive to odors. What is that sense organ?

Right! That sense organ is the nose.

You learned that light rays travel to the eye, and that sound waves travel to the ear. That's why we see and hear. Something like that happens with the nose too.

We smell because certain things travel to the nose. Those things are molecules. Molecules are very tiny pieces of a thing. The molecules float in the air. We breathe the air, and the molecules go into the nose. Then we smell the odor of the thing.

Look around you. Choose something that has several odors (for example, a sandwich or a plant with flowers). Smell it. What did you learn about the thing from the way it smells?

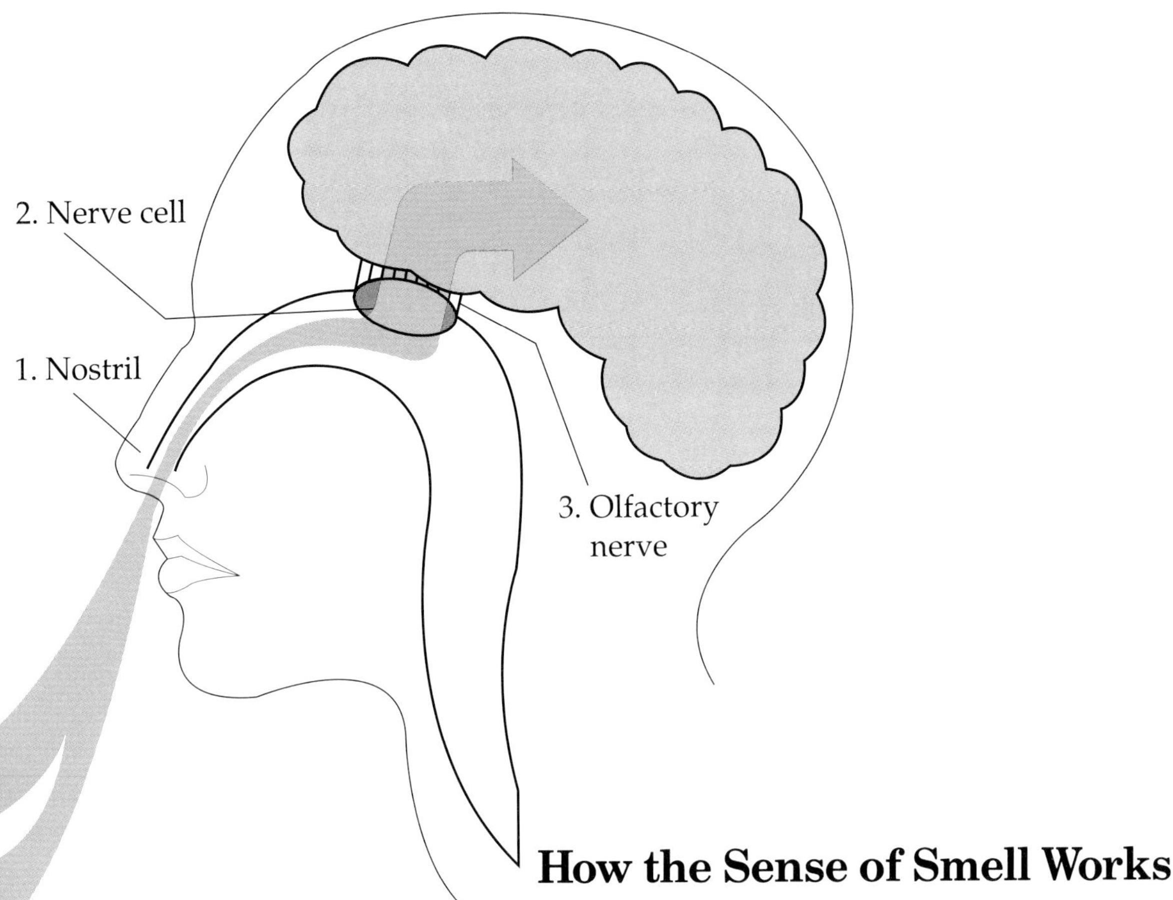

How the Sense of Smell Works

You learned that odors travel through the air to the nose. What happens when an odor reaches the nose? Look at the diagram. It shows the parts of the nose we use to smell.

The sentences below tell what happens when we smell an odor. But some of the words are missing. The missing words are the names of parts of the nose. You can find those names on the diagram. Find the words on the diagram that fill in the missing names in the sentences.

1. We breathe air through each ________. The air carries molecules from something that has an odor.
2. The air and the molecules go into the nose. They touch ________ ________ at the back of the nose.
3. The nerve cells send a message along the ________ ________.

The message goes to the brain, and we smell the odor.

Experiment 4

What happens when you smell an odor for a while?

Suppose you go to a hot-dog place. You have to wait in line for a long time.

You watch the hot dogs cooking on a grill. You hear the sound the hot dogs make as they cook. And you smell the hot dogs.

Finally, it's your turn to get your hot dog. You can still see the hot dogs cooking. You can still hear them cooking. What about the odor of the hot dogs? Do you think you can still smell it?

You see something for as long as the light rays come to your eyes. You hear something for as long as the sound waves come to your ears. Do you smell something for as long as the odor comes to your nose? Do this experiment and find out.

Materials

One watch with a second hand

Something that has a strong odor (such as after-shave or perfume)

Procedure

1. Hold the strong-smelling thing near your nose.
2. Smell it for three minutes. Use the watch to time yourself.

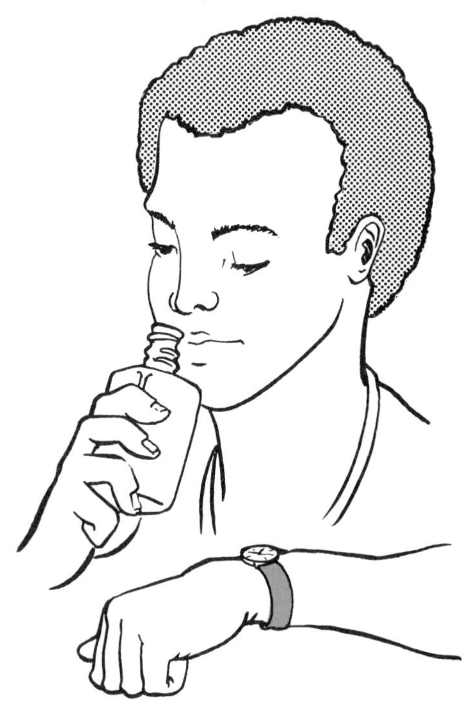

Observations

When you start, you smell the odor a lot. After three minutes, what do you smell?

1. You smell the odor more.
2. You smell the odor less.
3. You don't smell the odor at all.

After three minutes, you smell the odor less. Or you don't smell it at all.

Conclusions

1. What happens after you smell an odor for a while?
   a. You keep smelling the odor for as long as it comes to your nose.
   b. After a while, you stop smelling the odor that's coming to your nose.
2. Do you think it's a good thing or a bad thing that you stop smelling an odor after a while? Why?

Review

Use words from this unit to answer the questions. The words you'll need are listed below.

brain	molecules	odor
olfactory	nostrils	nose

1. What's another word for "smell"?
2. What sense organ do we use to smell?
3. What are the two holes in the nose called?
4. When we smell an odor, where does the message go from the olfactory nerves?
5. What carry odors in the air?
6. What nerves get the message from the nerve cells?

Check These Out

1. In some jobs, people are paid to use their sense of smell. Find out about some of those jobs.
2. Your sense of smell can tell you that you're in danger. Make a list of smells that may mean danger is near.
3. Some people can smell things that other people can't smell. Get a partner. Find out which things each of you can smell.
4. Everybody has some favorite odors. Make a list of your favorite odors. Tell why you like each one.
5. Here are more things you may want to find out:
   - How do animals use their sense of smell?
   - What is an insect's sense organ for smell? How does an insect use its sense of smell?
   - You know that you stop smelling an odor after a while. What happens if you smell something for a while, then take it away from your nose, and then put it back near your nose again?

Unit 5

The Sense of Taste

What are your favorite foods? Do you like hamburger? Fish? Chicken? Noodles? Apples?

One reason you like certain foods is that they taste good to you. Your sense of taste helps you know which foods you want to eat or don't want to eat. It helps you decide if a food is good or not good. It helps you decide if you want to eat more or stop eating.

- Why do we taste things?
- How does our sense of taste work?
- What kinds of things can we taste?

You'll find out in this unit.

Before You Start

You'll be using the science words below. Find out what they mean. Look them up in the Glossary. On a separate piece of paper, write what the words mean.

1. **dissolve**
2. **flavor**
3. **saliva**

Tasting Food

What is your sense organ for taste?

Right! Your sense organ for taste is your tongue. Your tongue is covered with nerve cells that are sensitive to tastes. Those nerve cells are called **taste buds**.

Get a mirror and look at your tongue. What is on the top of your tongue?

The top of your tongue is covered with many little bumps called **papillae**. Your taste buds are in the papillae.

When you eat a food, your taste buds send different messages about that food to your brain. Your brain puts all the messages together. Then the food tastes a certain way to you.

But something has to happen to the food before it can stimulate the taste buds. Find out what has to happen. Get a piece of hard cracker (or some other dry food, such as a sugar cube). Get a clean paper towel too.

Dry your tongue with the towel. Put the cracker (or sugar cube) on your tongue. Keep your tongue out of your mouth. Can you taste the food?

Put your tongue, with the food on it, in your mouth. Close your mouth and let the food become wet. Can you taste the food now?

You don't taste food when it is dry. You taste it when it becomes wet. That's because your taste buds are stimulated only by things that are dissolved in a liquid.

How does your saliva help you taste foods?

Four Tastes

Did you know this? Your taste buds probably are sensitive to only four tastes. Those tastes are *sweet, sour, bitter,* and *salty.* That's all!

Taste buds on different parts of your tongue are sensitive to those four tastes. For example, taste buds on the back of your tongue are sensitive to bitter tastes. And taste buds on the front sides of your tongue are sensitive to salty tastes.

You can find out which parts of your tongue are sensitive to sweet tastes and to sour tastes. You'll need something sweet, such as hard candy. And you'll need something sour, such as a slice of lemon.

1 Dip the candy into water to get it wet. Put the candy on the tip of your tongue. Now touch it to the two sides of your tongue. Where does it taste sweetest? Find that area on the diagram.

2 Now put the lemon on the tip of your tongue. Then touch it to the two sides. Where does it taste more sour? Find that area on the diagram.

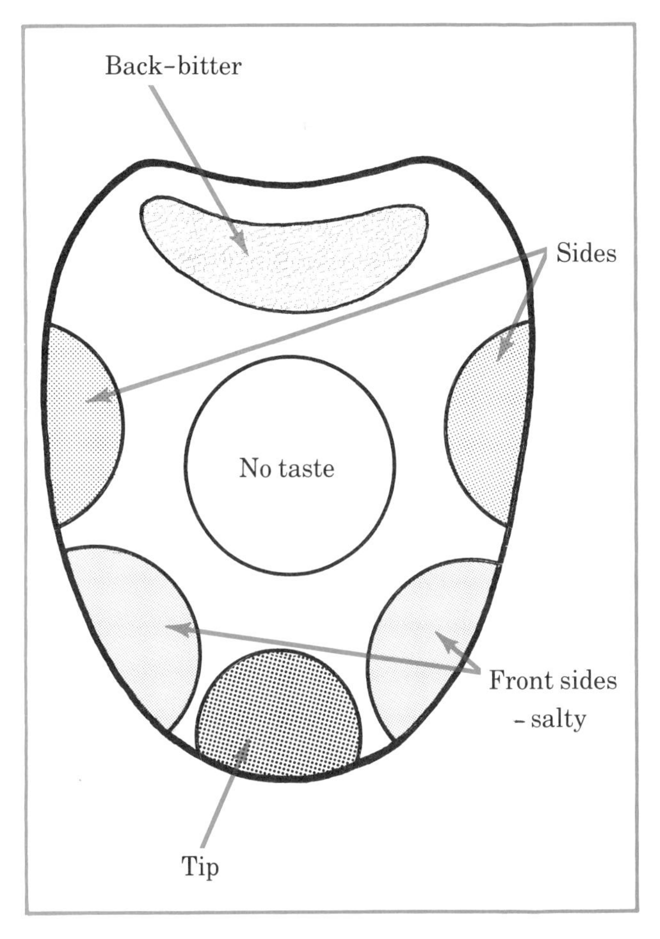

The taste areas on the tongue

What Happens?

1. Which part of the tongue is sensitive to sweet tastes?
2. Which part is sensitive to sour tastes?

(The answers are upside down.)

Answers

1. The tip is sensitive to sweet tastes.
2. The sides are sensitive to sour tastes.

Can You Smell It?

Let's say you have a cold. And you can't smell anything. You eat a hamburger. How does the hamburger taste?

The hamburger probably doesn't taste like anything when you have a cold. That's because food doesn't have flavor when you can't smell it.

You learned that taste buds are probably sensitive to just four tastes: sweet, sour, bitter, and salty. But we think foods have more than those four tastes because of our sense of smell.

Think about the way hamburgers taste. Hamburgers have a certain flavor, don't they? They don't taste like other food, such as chicken.

Here's why you think hamburgers taste the way they do:

When you chew a hamburger, some of it dissolves in your mouth. Your taste buds send messages to the brain about how sweet, sour, bitter, or salty the hamburger is.

But at the same time, your nose smells the odor of the hamburger. It sends messages about those odors to the brain. The brain puts together the messages from your taste buds and your nose. And you think the hamburger tastes like a hamburger!

Not all foods taste the way they do because of your sense of smell. But most foods do, like hamburgers. Think of some foods that taste good to you because of their odors.

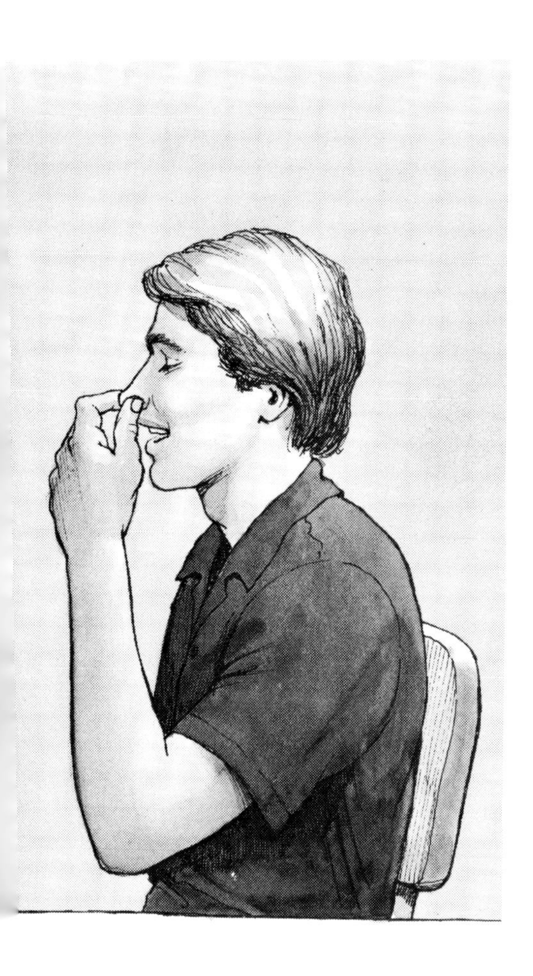

Can You Taste It?

How much does your sense of smell help you taste things? Find out. You'll need a partner. You'll also need small pieces of three different foods. (You could use apple, celery, onion, carrot, and so on.)

1 Close your eyes so that you can't see the foods. (Or put a blindfold over your eyes.)

Hold your nose.

Your partner now puts a piece of each food in your mouth, one piece at a time. Taste each food and tell your partner what you think the food is. Your partner then writes down on a piece of paper what the food is and what your guess is.

2 Now have your partner give you the foods to taste again. But this time don't hold your nose. (Keep your eyes closed.)

Your partner writes down what the food is and what your guess is.

What Happens?

When you *can't* smell, how many foods do you guess correctly?

When you *can* smell, how many foods do you guess correctly?

Review

Show what you learned in this unit. Finish these sentences. Match the first part of the sentence on the left with the correct words on the right.

1. The sense organ for taste
2. The top of the tongue is
3. You can taste only things that
4. Your taste buds are sensitive to
5. Each part of your tongue is sensitive
6. Most foods taste the way they do

a. is the tongue.
b. sweet, sour, salty, and bitter tastes.
c. covered with bumps called papillae.
d. are dissolved in water or saliva.
e. because of the way they smell.
f. to a certain taste.

Check These Out

1. Find pictures of foods that are sweet, sour, salty, and bitter. Make a poster out of those pictures. Label each food to show its strongest taste.
2. Have a "Tasting Day." Each person brings in a food that other people may not have tasted. It can be a family's favorite food or a food from another country.
3. What are food seasonings? Why do people use seasonings? Bring in some different kinds of seasonings.
4. Here are more things you may want to find out:
   - What does a taste bud look like?
   - What happens to dissolved food so that we can taste it?
   - What different shapes do papillae have? Do all papillae have the same number of taste buds?
   - The tongue senses other things besides tastes. It senses hot and cold. Why does a cold drink taste different from a warm drink?

Unit 6

The Sense of Touch

Close your eyes. Have a friend put something in your hand. Feel that thing. Even without seeing it, you can probably tell what that thing is just by touching it.

You learn a lot about your world through your sense of touch. Your sense of touch tells you if a thing is hot or cold. It tells you how heavy a thing is and whether it is smooth, hard, or sharp. It helps you find out what shape a thing is. It helps you know what a thing is.

- How does our sense of touch work?
- How do we use our sense of touch?

You'll find out in this unit.

Before You Start

You'll be using the science words below. Find out what they mean. Look them up in the Glossary. On a separate piece of paper, write what the words mean.

1. **pressure**
2. **reflex**
3. **sensation**

Our Sensitive Skin

You have a sense organ that completely covers your body. That organ tells you if something presses hard on you. It tells you if something as light as an ant touches you. It tells you how hot or cold things are and if those things hurt you.

What is that sense organ?

Right! That sense organ is your skin.

The skin is our sense organ for touch. It is so sensitive that blind people use it to read books. They do that by feeling special letters with their fingertips. (That way of reading is called **Braille**.)

The skin has millions of nerve cells that are sensitive to different kinds of touch. When something touches the skin, those nerve cells send their different touch messages to the brain. The brain tells us what we are feeling. We call each kind of feeling a sensation.

Think about the sensations you felt when you first put on your clothes today. Did your clothes feel heavy? Tight? Scratchy?

What sensations are you feeling right now? What do your clothes feel like?

You probably don't feel your clothes at all. That's because most of the skin's nerve cells stop sending messages after a while.

Why is it a good thing that these nerve cells stop sending touch messages after a while?

Different Kinds of Nerve Cells

The nerve cells in the skin are sensitive to *light touch, pressure, heat, cold,* and *pain*. But each nerve cell sends only one kind of touch message.

You can see that by making a map of the sensations that the hand feels. You'll need a partner. You'll also need these things:

- One piece of white paper
- Five colored pencils (black, yellow, red, blue, and green)
- One pin in a dish of hot water
- One pencil with a dull point
- One pin on an ice cube

Directions

Have your partner trace a hand on the piece of paper. Then ask your partner to close his or her eyes and to tell you when he or she feels a sensation.

1. **Heat**—Use the pin that is in hot water. Press the *head* of that pin gently all over your partner's hand. On the picture of the hand, draw black dots to show where your partner feels the heat.
2. **Light touch**—Very gently touch the point of the pencil all over your partner's hand. On the picture, draw yellow dots to show where your partner feels light touch.
3. **Pressure**—Press the point of the pencil a little harder on your partner's hand. Draw red dots on the picture.
4. **Cold**—Use the pin that is on the ice cube. Press the head of that pin gently on your partner's hand. Draw blue dots on the picture.
5. **Pain**—Gently press the point of the pin on your partner's hand. Draw green dots on the picture.

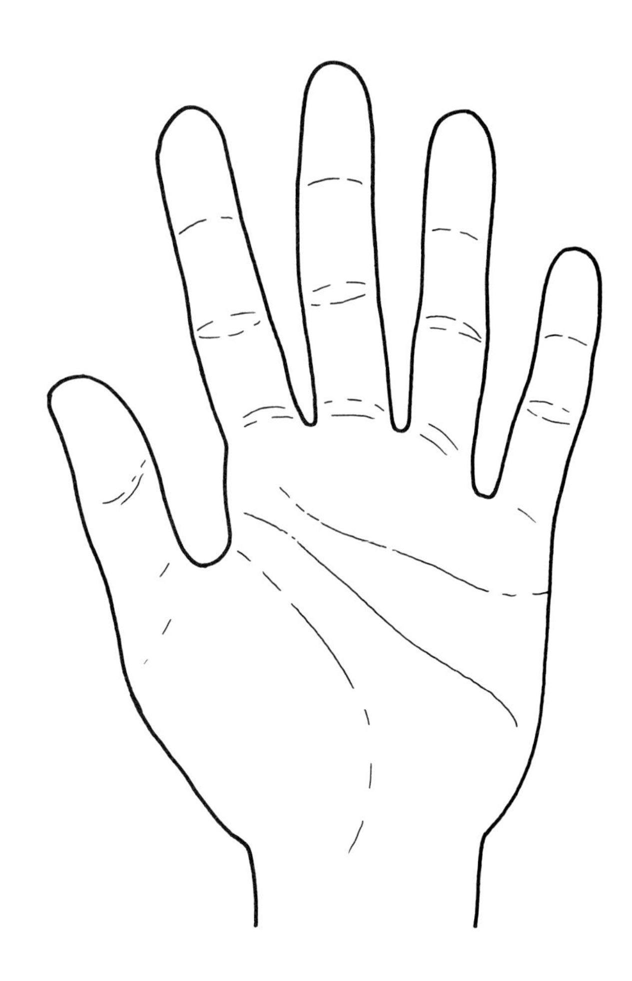

Experiment 5

Which part of the body is more sensitive?

Some parts of the body are more sensitive to light touch than other parts are. The skin on those sensitive parts has more nerve cells. And the nerve cells are closer together. So those sensitive parts feel more light touch sensations.

You can see that for yourself in this experiment. You'll need a partner.

Materials

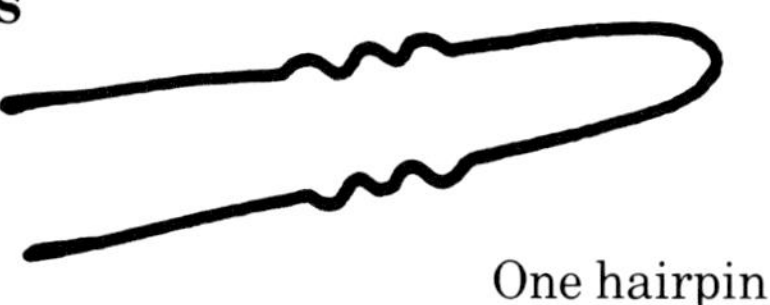

One hairpin

Procedure

1. Hold your hand, palm up, in front of you. Close your eyes.

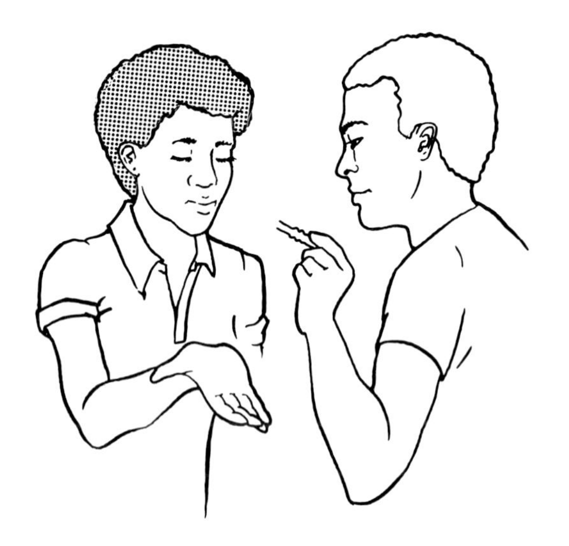

2. Your partner holds the hairpin so that the ends are about ½ an inch apart. Your partner lightly touches your fingertip with the two ends. You should feel two things touching you.

3. Your partner brings the ends of the hairpin closer together and touches your fingertip again. Your partner keeps doing that until you feel only one thing touching you. Open your eyes when you feel that. Look at the hairpin. How close together are the ends?

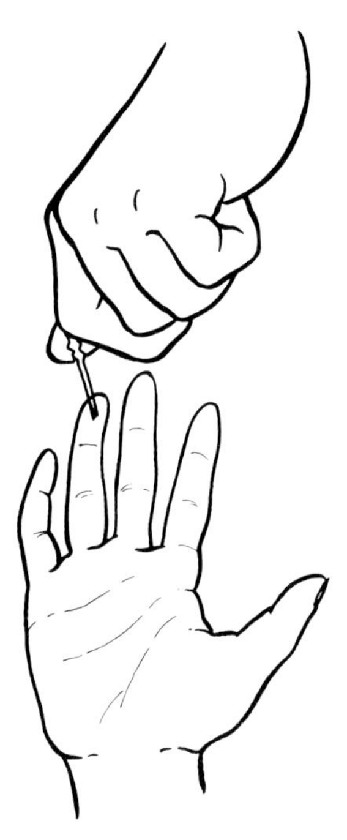

4. Close your eyes again. Your partner touches you in the same ways on your arm. Open your eyes when you feel only one thing. How close together are the hairpin ends?

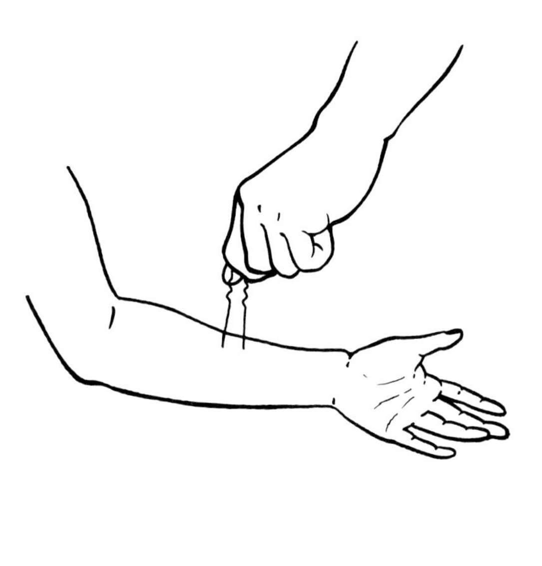

Observations

Where are the hairpin ends closer together when you open your eyes—on your arm or on your fingertip?

The hairpin ends are closer together on your fingertip. They are farther apart on your arm.

Conclusions

1. Your fingertip can feel two things when they are very close together. But your arms can't. Why is that?

   Right! The skin on your fingertip has more nerve cells than the skin on your arm.
2. Which part of the body is more sensitive: the fingertip or the arm?

Danger!

Suppose you're near a fire. Your skin feels how hot the fire is. So you stay far away from the fire, where you won't be burned. Your sense of touch keeps you safe.

But suppose you touch something very hot by accident. What does your body do?

The part of your body that touches the hot thing feels pain. That pain warns the body to do something fast. You quickly move away from the hot thing before it can hurt you any more.

In fact, you quickly move away from the hot thing *before* a message goes to your brain. That's because your body has a way of quickly acting when nerve cells feel certain kinds of pain. That quick action is called a *reflex*.

What is another example of a reflex?

Here's another way pain keeps us safe: Let's say you have a toothache. You go to the dentist. The dentist discovers an infected tooth. He or she treats the tooth, and the pain goes away.

A pain such as a toothache is a sign that something is wrong with the body. It can be a sign that you are sick. What would happen if you *didn't* have nerve cells that are sensitive to pain?

Review

Show what you learned in this unit. Find a word in the list below that fits each meaning.

pain	danger	sensation
heat	skin	light touch
reflex	cold	pressure

What's another word for

1. A kind of feeling?
2. A quick action?
3. What some nerve cells feel when you're in the sun?
4. What some nerve cells feel when you touch ice?
5. What some nerve cells feel when something presses hard?
6. A sign that tells you something is wrong?
7. What nerve cells keep you out of?
8. The sense organ for touch?
9. What some nerve cells feel when an ant touches you?

Check These Out

1. You know that blind people read by using Braille. Find out more about Braille. Get a Braille book and see what the letters are like.
2. Some people can tell what things are just by touching them. Get six different things. Choose a partner. See if your partner can guess what each thing is just by touching it.
3. Make a "touch picture." On poster paper, paste things that are big and small, rough and smooth.
4. Try doing some things using *just* your sense of touch. For example, try eating a meal, finding your way across a room, or finding a certain book.
5. To play some sports and games, people need to use their sense of touch. Find out more about those sports and games.
6. Here are more things you may want to find out:
   - What are the layers of the skin? What does each layer do?
   - What are the most sensitive parts of the body?
   - The skin is covered with fine hairs. What do those hairs have to do with the sense of touch?
   - What do the nerve cells in the skin look like?

Show What You Learned

What's the Answer?

1. Choose the correct sense organ to answer each question.

   tongue   nose   eye
   skin   ear

   a. What is the sense organ for sight?
   b. What is the sense organ for hearing?
   c. What is the sense organ for smell?
   d. What is the sense organ for taste?
   e. What is the sense organ for touch?
2. Choose one of the sense organs. On a separate piece of paper, tell how the sense organ gets information and sends it to the brain.
3. On a separate piece of paper, list the five senses. Give examples of how each one keeps us safe.

What's the Word?

Give the correct word or words for each meaning.

1. Part of the body that controls everything the body does
   B ________
2. Parts of the body that get facts about the world
   S ________ O ________
3. Parts of the body that are sensitive to sight, sound, smell, taste, and touch
   N ________ C ________
4. Parts of the body that take messages from the sense organs to the brain
   N ________
5. A kind of feeling
   S ________
6. Make something act or work
   S ________

Congratulations!

You've learned a lot about the five senses. You've learned

- What the five senses are
- How each sense works
- What your senses do
- And many other important facts about the five senses

Glossary
Green Plants

ab sorb To soak up water

car bon di ox ide A gas that leaves make food with

chlo ro phyll Something in a green leaf that traps light energy

dis solve To break down into very small pieces in water

en er gy The power that makes things change

fer til iz er A mixture of things plants need to stay healthy

flu o res cent A kind of electric light, usually shaped like a tube

gas es What air is made up of

leaves The parts of a plant that make food

light One thing plant leaves need to make food

main root The first root that grows out of a seed

nu tri ent Something plants need to stay healthy

ox y gen A gas that leaves give out

pho to syn the sis The way that green plants make food

pipe Something that water moves through

root The underground part of a plant

root hairs Tiny root parts that soak up water

sec ond ar y roots Roots that get water

seed ling A very young plant

soil The earth that plants grow in

sprout To start to grow

stalk A stem on a celery plant

stem The part of a plant that connects the root and the leaves

stored Kept to be used later

trans plant To dig up and plant again

trunk The main stem of a tree

wa ter One thing that plants need to grow

wa ter va por Water in the form of a gas

Glossary
Animals

bal ance To give all living things what they need

be have To act in a certain way

be hav ior A way of acting

bod y cav i ty The space inside a body

cir cu la to ry sys tem The parts in an animal that move food, oxygen, and other things throughout its body

class A group of animals such as birds, insects, and mammals

clas si fy To put into groups

de vel op To grow

di ges tive sys tem The parts in an animal that digest (break down) food

en dan gered To be in danger of dying out

en vi ron ment All the land and living things that make up a place

ex o skel e ton The hard covering on certain animals' bodies; an outside skeleton

ex tinct When one kind of animal dies out and none is left on Earth

fer til ize To join a male and a female sex cell and make a new life

in stinc tive be hav ior A way of acting that an animal is born with

learned be hav ior A way of acting that an animal is taught to do

life cy cle The different body changes an animal goes through from baby to adult

pred a tor An animal that kills other animals for food

prey An animal that is killed for food

re pro duce To produce new animals

sci en tif ic name The name that all scientists use to describe an animal

seg ments Small sections

spe cial ized Having special body parts; acting in special ways

spe cies A group of just one kind of animal; humans make up a species

stage A period of time in an animal's life

struc ture The shape and parts of an animal's body

sur vive To keep alive

sym bi o sis When two kinds of animals live together and help each other

ten ta cles Thin, moving parts that are used like arms

traits The certain ways an animal looks or acts

ver te brate The backbone inside an animal's body

warm -blood ed Being able to make and keep the same body heat

Glossary
Human Systems

ac tion A movement by any body part

cell The smallest living part of the body

cir cu la to ry sys tem The organs that move blood through the body

con tract To get shorter; to squeeze

di gest To break food down

di ges tive sys tem The organs that break food down

di vide to split apart

em bry o A living thing that's starting to develop its body parts

en do crine sys tem The organs that make chemicals that control the body

en zyme A chemical that helps break food down

ex crete To get rid of

ex cre to ry sys tem The organs that remove cell wastes from the body

fer til ize To join a sperm cell with an egg cell

fe tus An unborn baby that's in its mother's body

hor mone A chemical that makes cells do certain things

joint The place where two bones meet

lymph A fluid that carries waste from the cells

lym phat ic sys tem The organs that move lymph through the body

mes sa ges The facts sent back and forth between the brain and the body

mus cu lar sys tem The organs that move different body parts

ner vous sys tem The organs that control what the body does

nu tri ents The parts of food that cells use

or gan A body part that is made up of different tissues

ox y gen A gas that cells need to live

re lax To get longer

re pro duce To make a new living thing

re pro duc tive sys tem The organs that make new living things

res pi ra to ry sys tem The organs that move air and other gases in and out of the body

skel e tal sys tem The organs that give your body shape and support; the body's bones

skel e ton All the bones inside your body

tis sue A group of the same kind of cells

waste prod ucts Materials that your body makes when it uses food and oxygen

Glossary

The Five Senses

au di to ry nerve Nerve that carries sound messages to the brain

Braille A way blind people read

brain Part of the body that controls everything the body does

con nect ed Joined together

dis solve To melt in a liquid

ear drum A very thin piece of skin between the outer ear and the middle ear

en ter To go into something

eye lid Skin that closes over the eye

five sen ses Sight, hearing, smell, taste, and touch

fla vor A taste

fo cus To gather together

hear ing The sense that tells us about sounds

im age A picture

in for ma tion Facts about a thing

i ris The colored circle in the eye

lens Part of the eye that gathers light rays and helps us see things near and far away

mol e cules Very tiny pieces of a thing

nerve cells Parts of the body that get sights, sounds, smells, tastes, and touches

nerves Parts of the body that carry messages to the brain

nos trils The two holes in the nose

o dor A smell

ol fac to ry nerve Nerve that carries smell messages to the brain

op tic nerve The nerve that carries sight messages to the brain

pa pil lae Tiny bumps that cover the tongue

pres sure Pressing down

pu pil Hole in the eye that light rays go through

re flex Quick action because of something you sense

ret in a The back of the eyeball

sa li va The liquid in the mouth

sen sa tion A kind of feeling

sense or gans Parts of the body that get facts about the world; eyes, ears, nose, tongue, and skin

sen si tive Able to pick up light, sound, odor, touch, and taste

sight The sense that tells us what things look like

smell The sense that tells us what things smell like

sound waves The way that sound travels

stim u late To make something act or work

taste The sense that tells us if things are sweet, salty, sour, or bitter

taste buds Taste nerve cells

touch The sense that tells us what things feel like

vi brates Moves back and forth quickly

vi bra tion A very fast movement that goes back and forth